For Nicole + Carl + Cec'l - Elizabeth.
As a reminder of you ... d
to come again soon. ... time.

We Swept the Cornflakes Out the Door

by

Edith Hewson

The adventures of a prairie family

Dedicated to Franki and Doris and Geoff
with love and appreciation for all your help.

To be read individually & to each other!

Love, Mom + dad.

BRADLEY
PUBLICATIONS INC

Published by Bradley Publications Inc.
2330 15th Avenue
Regina, Saskatchewan S4P 1A2

Illustrations by Jack Gravenor and Doris Silcox
Cover design and illustration by Tom Boyko
Printed and bound in Canada by PrintWest Communications

ISBN 0-929035-13-5

WE SWEPT THE CORNFLAKES OUT THE DOOR

by

Edith Hewson

Katie Cameron McDougall's eyes glowed with missionary zeal.

I'm convinced", she wrote to her sister in Ontario, "that the demon rum could be beaten and the bars in Wapella closed for want of business if there were more young women of marriageable age here."

She paused, turned her head sideways to listen, then squinted nervously out the tiny parlor window.

An early chinook had decimated snowbanks. The raucous caw-cawing of crows echoed across the bright dome of prairie sky. Unplowed grasslands, warm brown and pale green, promising early spring, stretched to far off horizons. A cluster of toy-like red elevators was the only sign of habitation. A dusty trail lost itself over a rise that separated the unpainted frame house in which she worked from the prairie town of Wapella to the north. The twentieth century had just begun. Wapella was the centre of a rapidly widening wheat country, golden with promise that overflowed across the face of western Canada.

She bent again to her writing.

"These glorious plains, great tracts of land, can be had for a song and are just full of excellent opportunities . . ."

She paused and looked at what she had written.

"Well, they're excellent all right if you're not afraid to work."

The westering sun glinted on her rimless glasses and turned her mound of chestnut hair to burnished copper. Impatiently she scooped up strands that had sprung loose from her top-heavy pompadour and tucked them back in place. Her high-necked white blouse, stiff with starch, crackled each time she moved. A grey wool worsted skirt swirled about her black buttoned boot tops.

She glanced at the ad she had cut from the local paper on her arrival a month before. It read, "WANTED — NURSE COMPANION FOR SEMI-INVALID. CONVENIENT FARM HOME. EXCELLENT WORKING CONDITIONS FOR RIGHT PARTY. SEE J. BARING."

"Semi-invalid is a mis-nomer for the poor soul," wrote Kate. "She screams all night or croons to a battered old doll. During the day she crawls under her bed to sleep. She scratches and bites me when I try to get her to come out. I lock the door at night, otherwise she runs out on the prairie and hides." Kate paused, peered out the window, and smiled to herself.

"The convenient farm home must refer exclusively to the pump. It's right outside the door — instead of a mile away in the cow-pasture like the neighbours have it! I have to prime it every time I want water. You pour it in at the top and then pump like mad until a trickle dribbles out. It takes ages to get all I need — laundry, scrubbing, bathing, dishes — and the patient (she soils everything). Everyone says the water here is so good, especially for the teeth, but I can't see it. It tastes like epsom salts laced with carbolic! — difficult to lather even with my homemade soap, and it does nothing for your skin!" She stopped and looked down at her roughened red hands.

"I make the meals for Mr. Baring and the two hired men — all part of the job. There's no chance to read — nothing to read here anyway. So I sew, curtains and things. Made most of them out of dirty old flour sacks I found lying around." She paused as the evening breeze fluttered the snowy results of her handiwork.

"I guess what's meant by "excellent working conditions" in the ad, is you've an excellent chance to be working all the time."

Love to all,
Kate.

P.S. — Cousin Robert dropped by on his way to his church in the Scottish settlements south of here. He walks all the way. He says he preaches mostly in Gaelic. The older parishioners can't understand English yet. He worries because he coaxed me out here and says I work too hard. I told him I want to get you and mother out as soon as possible — and the boys too. Don't let this discourage you, Jessie, from coming out. There are lots of other jobs I could get but I can't leave the poor souls in the lurch. And besides, something's bound to turn up."

And something had. It was Sam Marks.

He was tough and stocky like his Cornish forebears. His big brown eyes, silky black hair parted in the middle, and curling moustache to match, appealed strongly to her when she first stepped off the train.

Women were rare and precious. The Wapella station platform was always lined at train time with lonely young men hoping to get a glimpse of a future likely bride.

"That Sam Marks' middle name should be Speed," volunteered Joe

Baring to Kate when her job was only twenty-four hours old. Sam had already come courting, made himself acquainted and gone.

"He never walks. Runs! Works two quarter sections and raises cattle besides. Never misses a harness race. Has the fastest team in the country. Wins most of the races, too. He's building a livery stable in Esketh. Drinks some." Baring shook his head doubtfully. "Everybody likes him, though," he added.

Sam Marks was a fast worker. Shortly after Kate's arrival he had shooed away all possible rivals and was on a first name basis with the Barings. Although he made no rash statements about renouncing his drinking habits, he had slyly helped Kate convince herself that the long, lonely winter nights suffered by prairie bachelors left them woefully unprepared to resist the temptations of bright lights and booze.

Kate wrote on rapidly, her ears attuned to sudden unexplained sounds from the patient's room above. Her myopic eyes squinted frequently down the prairie trail toward town. This day had been a hard one for her.

"The Missis is always worse when the moon's full," apologized Mr. Baring. "We'd best be on our guard tonight."

The singing sound of wheels and the rapid clip, clip of horses' hooves floated through the window. She rose hurriedly and put aside her pen. She flicked imaginary dust from the old organ and the table, gave her flowing skirt a yank and smoothed her rebellious hair.

A black bowler hat and the tips of two horses' heads that bobbed in unison, moved rapidly along the top of the hill. Moments later two perfectly matched black "drivers" pulling a new democrat swung into the yard. It was Sam. Soon he was seated in the spotless parlour opposite Kate, and Mr. Baring diplomatically withdrew to the kitchen.

Kate folded her hands and made the proper comments about the early spring weather with seeming serenity, but inside she was excited.

Sam, usually in control of every situation, was nervous.

"Kate, ——uumm, the winters out here are long and lonely,——" he began.

"Yes?" Kate's guileless query was punctuated with a thump on the stairs and the parlour door crashed open. Mrs. Baring, greying hair hanging in disarray to her waist, an old coat clutched over her nightgown, and in bare feet, burst into the room.

She raced to the organ, pulled out the stops and played several crashing chords. Next she grabbed the heavy glass lamp from the organ top, and screaming wildly, swung it at her nurse.

Sam jumped to intervene and caught the full force of the weapon across his arm. A wild tussle followed, the mad woman reaching for Kate's throat, pulling and tearing with maniac strength. Mr. Baring rushed in and for minutes the struggle went on. The two men at last held the madwoman by her arms. Kate, hair falling down and blouse torn, was valiantly doing her share to quell the melee.

When the woman had been soothed by a laudanum mixture prescribed by the local druggist, and safely locked in her room, Mr. Baring said, "I'm sorry, Miss McDougall, but the doctor says I must commit my wife to the mental hospital in Brandon."

Later, as a bright spring moon soared over the silent fields, Kate resumed her letter writing. Her face was scratched and red and her hair hung in strands; the starched white blouse missing two buttons was crushed, but her spirit was not.

"P.S." she added to her letter. "I'm getting married tomorrow."

CHAPTER 2

On the day following the altercation with the mad woman, Sam arrived with no warning at a friend's house in Wapella, with Kate and a preacher in tow.

"Mrs. Atcheson, put away your washtub," he announced, "there's going to be a wedding here today!"

His surprised hostess emptied her tub of soapy water, dried her hands, shoved the sorted and soiled clothes back in their basket and welcomed the exuberant bridegroom and his nervous bride. She took off her apron and put on her Sunday best. With a daughter and a neighbour as witnesses, Kate and Sam were married.

"Save your celebration," Sam said to the offer of a party. "Kate has to go back to the job, till Baring gets back from Brandon, and I have to go away on business."

They spent a short honeymoon at the local hotel. Then Sam left for Esketh where he finalized a business deal over his new livery stable.

Some days later, Kate. an unbridelike grimness about her chin, stared angrily into the looking-glass in the Atcheson bedroom adjoining the room in which she had been married. Guests were arriving, among them her bridegroom whom she hadn't seen for several days, and some of his former bachelor friends. Will, Sam's dreamy and absent-minded brother, was there too. He had ridden in with an invitation from his parents. They had just built a fine new log house and the bride and groom were expected to spend a few days with them.

Kate, comb in hand, was putting finishing touches to her unruly hair and had opened the door a crack to look at the crowd. She was just in time to hear the whispered remark — "Sam looks as if he had a skinfull last night —"

She snapped the door shut and stood in frozen anger, uncertain what to do.

"Drinking so soon after our wedding — after he promised —." Then she realized he hadn't promised anything and she was madder than ever.

What she hadn't heard was — "But he hasn't had a drink for a week. Looks awful pale, as if he's sick, but maybe he's just nervous."

"Bridegroom or no bridegroom, he's going to hear from me about this later." Kate opened the door and walked resolutely into the group that waited to meet her.

Sam stumbled at once to her side. His face was ashen. He swayed and grabbed for her hand to steady himself. She tried to shake off his clutch but he held on. He swung about and caught her around the waist. She struggled to push him to the vertical and he gamely tried to co-operate. For a moment or two, too long for the studied decorum of the times, they swayed, locked together in apparent unseemly ardour. The guests tittered nervously at the spectacle.

Kate, disengaged at last, gripped Sam's arm in a vice. Her pince-nez spectacles were askew and strands of untidy hair hung around her scarlet face as she turned him around to receive the well-wishes of the guests.

A bridal feast prepared by neighbouring women, all excellent cooks, was set before them. For ten years Sam, like most prairie bachelors, had lived on an unrelieved diet of bacon, beans and bannock. His bannock was a soggy flour and water mixture fried in salt pork fat, often burned black on the outside and doughy within.

Like an old warhorse revived at the first thump of the drum, Sam rallied at the sight of the banquet spread before him, but his revival was short.

He whispered to Kate, "Get me out of here — stomach ache — I'm going to be sick —"

Appearances meant a lot. Kate wasn't going to start her marriage embarrassed by the results of her new husband's hangover.

"Will," she whispered to her new brother-in-law, "get Sam's team out quickly!"

She piloted the weaving, white-faced groom towards the door, saying hurried but firm farewells as she went.

Outside, Will held the reins of the prancing team and half lifted, half shoved his brother into the driver's seat beside the bride. Then he passed the reins into Sam's slack hands.

"They do seem in rather a hurry," remarked Will innocently to no one in particular as he stepped back none too soon from the flying wheels of the democrat. With the other guests he watched with suppressed amusement as the fastest newly-wed takeoff in the west took off.

Kate, appearances uppermost, turned to wave a decorous good-bye. She had to grab the handhold instead, and her words were lost on the wind. The horses, sensing the loosened control, bolted for the road south, in a spray of stones from the rocking wheels. As they hit their race-track stride, Sam groaned, "Take the reins, Kate," and fainted.

The only associations city-bred Katie McDougall had with horses to date were a few slow buggy-rides with Sam or one of his several rivals during the first few weeks in the west. She grasped the lines in one hand and shoved her bridegroom forward between the seat and the dashboard, for safety.

"Wake up," she screamed. "I can't stop them!" But Sam was beyond caring. Her hat, swathed in a long veil and anchored at the back, flew off and swung wildly, kiting high into the wind, further terrifying the runaway horses. The democrat jounced over ruts, gopher holes and roadside rocks. She clung to the reins in rigid desperation as the horses leaped a ditch, caromed into a field, then back across the ditch onto the prairie trail. The last bounce revived her husband briefly. He retched over the side of the buggy and slipped again into oblivion.

A team and wagon dead centre of the road loomed ahead.

"Sam! Sam! Wake up! We'll both be killed!" Sam tried to rally but was almost pitched from the careening vehicle. Kate grabbed him in time and wedged him firmly down with her knee in his back. Then with a rein in each hand she grimly tugged to turn the horses' heads from the road. They swerved at last. The democrat again hit the bank, soared like a balloon and landed on spinning wheels with a bone-jarring thud. The farmer, half asleep on his wagon load of wheat, jolted awake as the team, foaming at the mouth and streaming sweat, pounded past.

"Sam Marks as usual going too fast. His horses all right — but who's driving?"

Before he had time to decide, the back of the bouncing democrat disappeared in a swirl of dust.

By the time they had covered a few more miles, Kate and the horses had come to an understanding. The entourage slowed and her husband revived. When they reached her new in-laws' farm, her emotions were a shambles. Joy that this was no alcoholic hangover was offset by terror that she might lose her new husband altogether by this mysterious malady.

When they got to the farm, Sam was very sick. Kate spent the night helping her mother-in-law apply "hot fomentations" to his aching stomach. The "fomentations" were cloths soaked in boiling water "as hot as the hands can bear" — the accepted home remedy for all body ills.

Will rode in late that night, bright-eyed and affable from an evening celebrating his brother's nuptials in a Wapella bar.

"O woman in our hours of ease
Uncertain, coy and hard to please!
When pain and anguish wring the brow
A ministering angel thou!"

he recited repeatedly in the middle of all the anxiety and confusion.

As dawn broke over the prairies, the democrat pulled out again, a fresh team in the traces and Will driving. The back seat had been removed and the sick man lay stretched on a mattress. His anguished bride knelt beside him, mopping his forehead, holding his hands and wiping her eyes.

"I'll never doubt you again," she vowed recklessly. "I'll never question anything you do!" (She had cause to regret these hasty vows in the first years of her marriage.)

Will, up front, clucked at the team and from time to time broke lustily into song —

"I'm a dead cowboy, I know I done wrong!" which Kate privately thought in very poor taste for the occasion.

A week later a Wapella doctor reported to Kate, "Brandon hospital just sent word that your husband is on the mend. They agree he has a lion's constitution.

Sam survived this attack of appendicitis, often fatal in those times, a wild forty-mile ride to and from the farm over rough prairie roads, and a burst appendix on the train to Brandon Hospital. This was wonder enough.

"But Mrs. Marks," warned the doctor gravely, "ice applied to the abdomen is the only safe thing to use on a person with appendicitis! Your husband's survival is a miracle indeed!"

CHAPTER 3

Sam's reputation for short cuts and speed had an early start. In 1886 when he was twelve, his father said: "Sam be uncommon quick in school.

'Ee can read and cipher too. 'Er bain't no use going to school longer. Eliot needs 'm for 'ayin'."

Sam joined the hired hands labouring in the hayfields of a large farm in Ingersoll county near Woodstock, Ontario. He earned his board and $4.00 a month, $2.00 of which went into the family sugar bowl. His parents, landed immigrants from Cornwall, England, worked out nearby. Their earnings were small and life differed little from the grinding drudgery of their tenant-farmering days in the old country.

"You work well, son," remarked Eliot almost a year later. "I'm payin you $6.00 a month this year if you can keep up your speed." The flattery was lost on the boy. Two men less were hired for haying in the years that followed.

One hot August evening Sam trudged wearily home, his meagre belongings in a bundle on his back. He was a few months short of being sixteen. Startled parents greeted this unexpected visit with some dismay.

"I'm going out west with the harvesters," he stated firmly.

"But 'ee be well liked by Eliot." sputtered his father. " 'Er by saying' 'ee be very quick — and a good worker too 'er sez!"

Sam's father, fearful of parting with his eldest son of whom he was most proud, was also thinking of the loss to the family sugar bowl.

"Eliot says 'ee be good with 'orses too, an' can tame any one by gentlin'! Dreadful uncanny 'er be sayin' about 'ee!"

"Father, we'll never get rich workin' for Eliot. I'm going west and get homesteads for us all. We can be our own bosses then. Brother Will can have his own farm. Don't let him work for Eliot. It's too hard. Keep him in school."

The next morning, with his father's weakening objections and his mother's nervous prayers ringing in his ears, Sam clambered onto the excursion train headed west. A big pack of belongings strapped to a work-stunted, tough, little body and a look of stubborn determination on a brown, still-unshaven boy-face was the last his family saw of him for some years.

In the interval he earned a man's wages, harvesting, working in livery barns and riding herd on a Yorkton ranch in the bitterest winter weather. He acquired some cattle, horses and a homestead when he was legally well under age to do so. To prove up his land as required by law, he built a fine big log barn and good outbuildings for stock. When it got too cold to sleep on the ground or in a tent he built a shack or "Soddie" for himself.

"Crocuses cover the prairies," he wrote to his family. "Blue and white wood violets are commencing to peep out in the woods around. Open land for homesteads abounds. My house is snug in winter and cool in summer and there's a fresh water spring —"

Prairie fires that swept across the plains in the early days were a constant threat to the new settlers. Fireguards, several rows of plowing, circled every farmstead. The slices of new prairie sod, held together by tough grass roots turned up by the blade of Sam's breaking plow to protect his precious barns and stock pens, were also the bricks he used to build the house.

One small window in the back wall and one rough built doorway in front let in the only light. Little was needed. He spent most of his waking hours hurrying from one farm task to another, outside.

In winter when nights were long, his young bachelor neighbours came to share their loneliness. They brought fiddles, flutes and concertinas to accompany Sam's mouth organ. Thin orange-wrapper tissue paper over a comb and hummed through, worked well too. Firelight glowed from the top of the squat iron cookstove in the centre of the room. A candle or two stuck in the sod walls cast light enough while they made music, talked "get-rich-quick" schemes, or snored off the results of a shared bottle of home-brew or store-bought whiskey.

The roof of the shack was a row of poplar poles, raised in the middle to shed the rain, and thatched with slabs of turf. In spring the sods burst into life. Swallows, sparrows and wrens nested under the poplar poles. Bluebells, buffalo beans and dandelions waved and bobbed about in the grasses that sprang from the roof like the furbelows and flounces on a lady's Sunday hat. Inside the floor was solid prairie gumbo, hardened and polished by the passage of many feet.

Sabbath morning was usually cleaning day. Sam made a broom of red willow withes wired to a thin poplar pole and swept the week's accumulation of clods of mud, empty cans, old papers, bottles and potato peelings into the yard. As the seasons passed the accumulation became richer and higher, a fine breeding place for flies and a cosy home for field mice that became house mice as the weather worsened.

Sam convinced his family that they should join him. They arrived one sunny summer day in a wagon outfit bought in Wapella. The wagon was piled high with woven straw trunks, boxes and barrels of household goods.

Sam, now a tough-muscled, black-mustached man, bounded out of the shack to greet them, surrounded by the usual complement of a bachelor's company in those days — a yapping, barking pack of mongrel dogs.

His mother, dressed in a tiny-waisted black taffeta coat and grey plaid dress that trailed the ground, was helped from the wagon seat. Her winged hat soared over the pompadour of silky black hair and made her appear to tower over her fierce little Cornish husband.

She surveyed the large log stable built cosily against a grove of graceful aspens and the gnarled little mud hut encircled in garbage, with mixed amusement and pride for her son's work.

"Might we trade houses with thy stock, Sam?" she inquired gently. "They be better housed than thee, I be thinkin' ".

"Ee don't 'ev no floor in't," yelled his father after a rapid inspection of the "soddie" inside and out. "H'it be 'avin' two pipe 'oles and only one stove pipe." He scurried in and out banging the narrow wooden door that swung from leather hinges, yelling out one deficiency after another.

His wife waited serenely till he was run down. A religious woman, she stopped in silent prayer just inside the dingy shack door and her husband toned down his scolding one octave in deference to her.

He trotted about the yard till he found the poorly hidden hole in which his son had half-buried some liquor bottles.

"Ee be drinkin' some then." He looked half-approvingly at the bottles and said in a loud whisper, "Mmm, good stuff, eh? See ee always drinks the best."

Will took the team to the spring to drink and leaned dreamily against a post watching the birds twittering about the stable eaves. Soon the horses pulled away and began to crop grass nearby.

"Gor' dairney all," snorted his father. "Will! Will! Look sharp. Bain't ee see the 'orses wanderin'?"

He turned to Sam with a half grin. "Will don't talk much but ee be a beggar to think."

This arrival of the Marks family in western Canada was the culmination of a dream that began in 1874. Sam was born in Horsepool Cottage across a gully from the miniscule village of St. Mellion, Cornwall, England. When his parents gazed on the face of their tiny first-born son, they resolved the limited life of tenant farmers was not enough. They would go to Canada where land was free and opportunity for a better life greater.

Sam was christened in the ancient St. Mellion Church, built in 1259. His mother, deeply religious by nature, was attracted to and later joined the Methodist Church in the neighbouring village of St. Dominic. Her decision was not wholly endorsed by her family or her earthy husband.

The couple had another son, William. They had worked hard and by 1882 had saved enough from the sale of their garden produce on market days to pay passage to Canada.

On the day of their departure, gusts of spring rain beat against the windows of Horsepool Cottage. Relatives, Rickards and Markses, jammed the tiny rooms and crowded about the departing foursome. The husband, dressed in natty tweed and deerstalker hat, tapped his cane impatiently. His wife, in the sombre colours of the Puritans, and the two small boys in sturdy homespun, were in turn fearful and jubilant in the centre of all the excitement. Prayers were intoned and hymns sung intermittently, interspersed with loud sobbing and exclamations of regret by those being left behind.

Suddenly a tall, spare woman, hair as black as a raven's wing and little black eyes fiercely bright, appeared from the rooms above. In one bony hand she grasped a stone jug of Cornish mead. She made her way slowly to the centre of the room and viewed the noisy gathering with seeming disgust.

"More damp in nor rain out!" she croaked. She seated herself by the hearth and the din quietened. "Come, grandsons, to your grannie."

Sam and Will pressed close to her side and she put the jug on the floor. She put her arms around the little boys.

"Bring three mugs," she demanded.

She glanced defiantly at her daughter, filled the mugs from the jug and placed one each in the little boys' hands.

"We'll drink now," she intoned. "I'll get no chance to drink wi' ee when grown." She pressed the boys to her.

"And may God 'old ee in the 'ollow of 'is 'ands! Drink up, lads!" The fierce old lady and the two tiny boys drained the mugs while the assembly looked on in silence.

With this benison following them, they embarked on the vessel, the Constitution. The ship, which had been converted recently to steam, luckily had the sails still furled.

"Your great-grandmother's blessing worked," said Sam, telling it later to his children. "I escaped death by a miracle, and the vessel had been given up for lost when it limped into Montreal harbour weeks overdue."

Two days out from England, a storm of hurricane proportions arose. Sam's luckless parents, with a crowd of other immigrants, were sealed below decks, all hatches battened down to keep out the raging waves. The engine, damaged by salt water, was useless. A broken rudder added to their misfortunes, and left the ship to the mercies of a merciless sea.

Their tiny stateroom, with two narrow bunks to be shared by four, had standing-room only. It looked clean and inviting when they stepped on board. Days later Sam's mother lay helpless in one bunk, and her husband not much better in the other. Their sea-sickness was worsened by the fetid air from the packed humanity sharing the same unhappy experiences in the bottom of the pitching, lurching ship.

The captain had warned passengers to make their peace with God, for no adult on board expected the ship to reach Canadian shores. The children, with the resilience of the young, suffered little from sickness. They played and raced about, oblivious to the fate that waited in the lashing seas surrounding the stricken ship.

The storm lessened at last. A sailor went up a ladder and left the hatch slightly ajar. With a rope about his waist he fought his way along the slippery deck with great difficulty, inching hand over hand along the railing. Sam, followed by Will, crept up the ladder for a peek at the swirling world above, their first glimpse for many days.

"Go ahead, Sam," urged his brother. "I can't see ought."

"They say we bain't go up. We bain't."

"We be but little fellows," whispered Will. He gave his brother an encouraging push from below. "They won't notice ought." And Sam moved gingerly out on the heaving deck.

At that exact moment the beleaguered ship lurched sideways as a gigantic wave crashed over the gunwales, catching the tiny boy up like a cork. Will was astounded to see his brother's feet above him suddenly disappearing in a sheet of boiling green sea water.

The sailor returning along the deck was, with great difficulty, working his way back to the opened hatch. He was just in time to see the wave strike Sam. He let go the railing, and in a flying tackle, grabbed the boy's feet before he could be swept overboard. They both crashed into the railing as the sizzling waters rolled harmlessly back into the sea. Will ducked below and sat frozen with shock on the bottom rung until the sailor with his half-drowned and terrified brother appeared above him.

When the storm died the sails were unfurled and hoisted. As they disembarked in Montreal, every passenger, and particularly Sam Marks, knew in their hearts that being alive was a real miracle.

For ten years the family lived in Woodstock, until Sam made the break and convinced his family their future lay on the western plains too.

He helped his father and brother to establish homesteads on neighbouring land. He was busy, and women were scarce. Not until he was twenty-nine did he meet and resolve to marry Katie Cameron McDougall.

CHAPTER 4

Several weeks after their wedding in Wapella, Kate collected Sam's sparse supply of kitchen utensils and packed her one shabby bag. She skewered on a large feathered hat, bade her new in-laws a warm farewell and drove off to Esketh with a fully recovered husband.

For miles they clipped along a trail on the edge of the Qu'Appelle. Fresh green fields rippled in the valley below. Meadowlarks trilled from fenceposts. The rhythmic hum of the wheels made happy music and Kate's heart sang. Sam, beside her, drove with his usual racetrack enthusiasm.

"What's the house like?" She grabbed for her hat and stifled a gasp. The team, never slowing their pace, executed a tight curve with only a slight lift of the wheels on one side.

"It's all paid for! Got it dirt cheap! Don't owe a nickel on it!" Sam neglected to tell her he had won it in a poker game from a remittance man who was the town's worst drunk.

"I just can't believe it! A home of our own! All paid for!" Kate stole a proud, sidelong glance at her husband. She laughed joyously and hung on when the team navigated the next curve.

"Has it a view? I love a view!"

Sam looked thoughtful. "Town's growing fast. Lots of building going on. Great for resale value. You'll like it. Near the barn, too." He snapped the whip over the horses' heads and changed the subject.

When they reached the town limits the team swung onto a muddy rutted road. It was lined on both sides with dingy one-roomed shacks. They slowed and stopped outside the dingiest. Sam leapt over the wheel, tied the team to the sagging gate post and shifted boxes and bags inside the door. He didn't look at Kate's face.

"I'll be back in an hour." He untied the horses, jumped in the rig and was off down the street in the gathering gloom.

Kate stood in the centre of the room, surrounded by their few belongings. A light board partition separated a place for a bed on one side. Out-dated calendars hung drunkenly from every dirty wall. Old papers and garbage littered the gummy floors. The stale odour of mouldy cans and dried vomit was everywhere. A scrawny yellow tomcat sidled in and meowed expectantly over an empty tin dish.

Kate wept.

An hour later her hair hung in wet strands about her flushed face. Her skirt, hitched up behind with a large safety pin, remained immaculate through a fury of cleaning and scouring. She swabbed the floor with soapsuds well laced with carbolic and lye, scrubbing each patch vigorously with a wornout old broom. Water was replenished from an overflowing rain barrel outside. The results were beginning to cheer her.

"After all, we are just starting up." She thought of the wedding supper and how mistaken she had been over the cause of Sam's illness.

"I almost lost him over it. I said I'd never complain again and I won't." She straightened up and surveyed the improvement in her surroundings with pride.

"Well, it's all paid for. We don't owe a cent on it, and something's bound to turn up." She tossed out the last pail of filthy water and went to straighten her hair and don a clean apron.

Hours later Sam, fortified with several double whiskies, opened the sagging gate. Mounds of garbage ringed the front step. He carefully unlatched the door. The clean smell of soap and carbolic rose from the damp floorboards. Fire crackled in the cookstove, which was now polished and shining with black lead. An iron kettle bubbled on the back. The tomcat, paws folded and eyes closed, purred contentedly beneath the stove. Sam tip-toed in, removed his outer clothes and slipped into bed beside his sleeping wife.

Their living accommodations now differed only in kind, not in size, from Sam's bachelor home. The shack was the exact dimensions of the homestead soddie, and it didn't improve as the months flew by. It was freezing in winter and stifling all summer. No water handy; no lights; no cellar for storing things; but it did have lots of stovepipe holes.

"Sam's a genius at shortcuts," wrote Kate to her family. "To accommodate the cradle for his first-born he removed a little bedroom heater. Then he arranged the cookstove pipes through more holes cut in the dividing wall in a coming-and-going-pretzel-like arrangement. It saves room, distributes heat, and eliminates a lot of woodsplitting. It wouldn't win a prize for home decorating, but it does work!"

Her purpose in coming west was to get land and bring her family out to share it. Living in a cheerless shack on the edge of a brawling frontier town was away off target.

"I do want you to come out, Mother," she wrote. "We're bound to have better accommodation soon. Sam's business is improving and we don't owe a cent."

But Sam was happy. In place of burned bannock, dirty dishes and the chill of a bachelor's lot he came home to good food, cleanliness and warmth. Silent lonely nights on a raw homestead were exchanged for the bustling excitement of a new town in the making, and jovial evenings spent around the local hotel looking for business.

"I'll be away to Cannington races for a few days, Kate," he announced one morning. "I've got a chance to enter this horse I bought last week. Give me some free advertising if I win. I have a fare to take there anyway — kill two birds with one stone." He grabbed his straight razor and an extra shirt and threw them in a bag.

"But who will run the livery while you're gone?" Kate looked shocked at the prospect of losing business over a horse race.

"I hired young Jones to take over while I'm away. I'm going to keep him if he proves out well. Need an extra man or two around the barns anyway."

"Barns?" queried Kate. "Barns?"

"Carpenter's starting an addition or two this week. We'll need them with business expanding so fast." He was gone out the door before Kate could think of an answer.

"New barns? A stable boy? And more horses! Promise or no promise I intend to give him a piece of my mind when he gets back. I won't stay here another winter!" She slammed the door after him. The only result was that baby Mary, already demanding attention, howled louder.

On Sam's arrival home, late one night, there was little chance to broach the subject of her discontent.

"I won five races," he shouted at the door. "Did a little trading too. Bought two farms along the valley rim. Good land and excellent water on both. Should bring in good rent. Got people interested in one already."

"Couldn't we move out ——," Kate began. But Sam didn't hear.

"Had a great time at Cannington. They've imported the best horse-flesh on the continent. This team I bought have great track records and are harness trained too. Good ad for the business. It's been a great year." He was in bed and snoring comfortably before Kate had a chance to talk about moving, nor did she get much chance in the busy days ahead.

Sam leapt out of bed each morning and was gone most of the day. He rushed in to wolf his meals and arrived late at night to sleep. His reputation for speed was made. It didn't hurt his business one whit when the travellers he carried arrived shaken and bug-eyed at their destination. It was a challenge of nerve to ride with him, and for a time his repeat business was staggering.

This galloping success received a sudden jolt.

The bulk of the settlers in the town and the surrounding area were all from the same district in the old country.

"Funny lot," summed up Sam. "They tell me they couldn't snare a rabbit to feed their hungry families over there. Their landlord, a titled fellow, couldn't stand the sight of them starving so he got the Canadian government to take them off his hands. They were bundled onto immigrant ships and when they got here they were given great chunks of this open prairie to farm the best way they knew how. They are making a good job of it, too. But now they brag about this landlord who horse-whipped them and starved them out, as if he was a king or something. They're all good workers, I'll say that for them. Decent people, too, and honest ——"

Habits ground into these pioneers by generations of tyrants in the old land died hard. A new church head from their former homeland arrived in town. He was a thin, ferrety morsel of a man with a sharp, ugly face and a disposition to match. His close-set eyes glittering under the brim of his flat black hat struck members of the populace dumb with a nerveless fear, from old habit.

"To Tantallon, driver," he ordered one morning. He stepped into the back of Sam's livery with the air of an overlord commanding a slave. They spent the day going from one poor homestead to the next. Sam's overtures to conversation were ignored. "Drive here," or "Stop over there," were the only words spoken.

At each stop the cleric came out of the shack with a pleased smile on his face and money in his hands. Silver was deposited in pockets about his person, and bills were added to a growing wad he put into the bulging leather pouch on a belt about his waist.

"He's gouging money out of every one of those poor sods," Sam muttered. "And they're afraid of him, too — What for, I wonder?"

Back at Esketh the passenger clambered down from his back perch, and with a curt nod he was off down the street.

"Hey, where's my money?" Sam shouted after him.

The little man turned, weasel eyes blazing. "No money!" he spat out.

"you pay me or I'll have the law on you!" Sam threatened, still stinging with the insults he had taken all morning.

"You give ME money, or you be sorry."

After some argument, his rat-like face red with fury, the cleric threw some silver in the bottom of the democrat.

"I see you get no business. You not give me money — you leave!"

It was soon evident this was no idle threat. Local business took a sudden slump. Men who formerly sought Sam's advice or company seemed distant and hired other means of transport. Companionship at the local bar now assumed a larger importance in his life. Travellers and persons passing through had no religious differences when it came to sharing a drink.

"Got a cash offer for the livery stable tonight — fella from the east." It was 2 a.m. and Sam had just rolled in, bright-eyed and expansive. Kate, giving baby Mary her night feeding, looked startled.

"He was looking for farmland, but I told him farm living here was pretty hard on a woman — too many conveniences to give up."

"What conveniences?" snorted Kate. Before she could put the baby to bed, Sam was snoring, a smug smile on his face.

Mid-morning the next day he rushed in for his usual quick bite of hot biscuits and tea. "Going to Hazelcliffe to look at one of the farms. Fella interested in renting it."

"I'm going too, Sam."

"Can't wait for you. Too long a trip for the baby anyway ——" But the words faded on his lips unheard. Kate whisked diapers off a line stretched across the room, popped them into a basket where the baby slept, and disappeared out the door. She put the basket on the back seat of the democrat, rolled the baby into a shawl, and was pinning on her hat by the time her husband had swallowed his surprise along with the hot biscuits and tea.

Kate sat bolt upright in the back seat, kne arm clutching the baby and the other hanging onto her hat. She listened avidly to the sales talk her husband gave his fare as they spun along the winding trail on the rim of the Qu'Appelle valley.

"Look at that view!" Sam enthused, waving the buggy whip towards the valley floor that spread out beneath them. Clumps of aspen and patches of prairie oak glowed fiery red, orange and green. Chokecherry bushes and wild high-bush cranberries hung with purple and scarlet fruit along the way. Piles of whitened buffalo bones lay scattered by the bushes below. The river glittered like a cobalt snake between the quiet hills that waited for winter on this warm prairie autumn day.

"Great grazing land there — and the back quarter's good for forty bushel to the acre!" Sam, with the aplomb of a super-salesman, showed the worst farm first. The easterner was not impressed with the tiny four-roomed house and the outbuildings scattered about near a clump of trees. Then Sam touched up the team with his whip and they spun back along the valley's rim.

The team slowed and pulled into a houseyard set on the hill overlooking the village of Hazelcliffe. A red elevator, a church spire and frame and log houses lay scattered among the trees below.

A neat log house against a background of flaming maples on the property looked very inviting.

"Nice roomy place, this," Sam enthused, unlocking the door. "Kitchen window looks right over the valley. Nice view for the wife." Kate followed from room to room with the baby still clutched in her arms.

"Two big rooms here and more rooms upstairs. Good water, too — well right outside the kitchen. Logs all chinked and tight as a drum." He tapped the logs with the buggy whip handle.

Kate thought of the cramped box in which they lived, the town pump two ugly streets away, and the desperate struggle it was to save soft water when it rained.

"Good barns — log too, but sound as a nut ——" Sam's voice trailed on during the return ride. When they reached home he disappeared for the rest of the day, and Kate made her plans.

In the morning before leaving for work, Sam did his bookkeeping. Kate washed the dishes.

"That home on the farm above the village would be nice to have," she began.

"Mmmm," her husband continued writing.

"Buildings are good, and with the school just half a mile away ——"

"Uh-hhh——"

"It would be a good place to raise children — the next one ——"

"Sure hope he likes it — couldn't be handier for his family," and gathering up his accounts, Sam was gone out the door.

Kate gritted her teeth, picked up a plate, swung it in indecision and biting her lip to keep from swearing, she dashed the plate on the cabin floor where it splintered across the boards.

Next day she tried again.

"That place would be just right for us," she began. "A lovely view of the valley below, and the water so handy ——"

"Mmmm," murmured Sam. "Hope his Missis thinks so too. Took them both out to see it again today."

Kate, open-mouthed, stood for a moment and glared. "Damn!" she yelled. "Damn! Damn! Damn!" She stalked into the bedroom and swung the flimsy door shut with such force that every shelf in the shack shook.

Disturbed by his wife's unexpected profanity, Sam shook his head at the vagaries of women. He was soon off to work, whistling as he went.

The nights got colder and for Kate — alone every night — longer. In five months, her next child was due. Hopelessly she surveyed the crowded shack and wondered what to do for room.

"Tonight I'll tell him right out — soon as he gets home — that we've got to move. That farm overlooking the valley would be just great. We'd be further from the hotels, too ——"

Kate made trip after staggering trip to the town pump with a heavy baby on one hip and water pail in the other hand. Working the handle of the pump was difficult with the baby on one arm. Today she had to stand in line while others used the pump ahead of her. The baby, tired and uncomfortable, howled its displeasure. Sam didn't come home all day.

At night, she put the baby to bed, settled herself to wait and sat down to read. The clock ticked on, hour after hour. The fire died. Cold seeped in. As the cabin cooled, Kate's temper flared. She would stand it no longer.

She poured water on the last of the embers in the stove, and blew out the lamp. Then, covering the baby's cradle and grabbing her coat, she stepped out into the icy darkness.

A light snow fell and the streets were dark and quiet, with one exception. From the town hotel some distance away the muffled sounds of gaiety rose and fell. She hurried purposefully towards the sound. The big windows glowed warmly behind half-shuttered blinds. Men's laughter and the clink of glassware floated out on the silent street. She stopped among the rubble of a new building nearby, and listened. Behind those windows Sam, with his cronies, was laughing and drinking at the bar in clouds of cigar smoke.

She bent over, picked up a brick, hefted it and let fly at the plate glass windows. There was a lovely sound of splintering glass as the huge window buckled inward. The sudden explosion of silence from within hit the night air. She dusted off her hands, turned, and walked calmly home and went to bed.

Sam arrived home shortly after, and lit the lamp. Kate was innocently asleep. Beside the bed her shoes sat in a widening pool of melted snow.

"I've sold the livery stable and one of the farms," Sam announced the next afternoon. "How would you like to move out to the house on the farm above Hazelcliffe?"

CHAPTER 5

The move to the roomy log house above the village of Hazelcliffe was made at once. Willie was born in the spring, and Ruth the next year in the fall. The summer following, Kate's battle with the demon rum was a success. Sam quit drinking.

The farm took much of his time; Esketh was some distance away, the village of Hazelcliffe had no bar. One night he attended a religious revival meeting. The next day he threw away all tobacco, and he never took another drink.

"Do come over soon," Kate wrote to her mother-in-law. "I've wonderful news for you! Sam's turned over a new leaf ——"

The leaf turning didn't change his impatience with hired help. Sam moved fast himself and expected others to do the same. Like his father, he was a yeller too. A shouted order across the fields could be heard for miles on a clear day.

"You stupid jackass!" he yelled at a shy, young, city-bred man who was trying to learn the intricacies of farming. "What did you do with the bolt?" Sam found the bolt himself, but neglected to tell the unfortunate man until the next thing got lost. Despite the yelling, hired men never wanted to leave when their time was up.

At threshing time, Sam was more impatient than ever. Over the newly-installed telephone, he was rounding up "harvesters" just arrived from the east.

"'ow many did you say?" he yelled. He always yelled on the phone, as if his voice was carried by wind instead of wire.

"I'll take four of them, but don't send 'heny of those stupid greenhorn 'Hinglishmen!"

"Listen to you!" Kate laughed. "Can't stand your own countrymen!"

She was secretly pleased, for she felt the Scotch were just a cut or two above most English.

This antipathy for the English greenhorn dated from the Cecil Privett affair. Cecil's Cornish family happily rid themselves of him by a unanimous decision. They clubbed together and shared his one-way travel expenses to Saskatchewan. He was pushed on a boat at Plymouth with his satchel in one hand and the Marks' address in the other.

"There's a new man coming from England." Mary, the eldest child, followed Sally, the new hired girl, from garden to kitchen to sink.

"Maybe he'd make a good husband!" Mary watched all of Sally's romances with the intense interest of an imaginative matchmaker.

"Mmmpp." Sally pretended disinterest, but she took special care with the bread-making that day, and washed her face in buttermilk before she went to bed.

Sam picked up the new man in town. When Cecil stepped from the democrat Sally's eyes flew open in disbelief. She turned her back on the newcomer and spoke not a word to him for the painful duration of his stay. His resounding enthusiasm for her cooking failed to change her attitude.

Cecil's appetite was voracious. He snapped and gobbled at everything on the table with the manners of a half-starved wolf, swallowing chunks of meat, hunks of bread and forcing it all down past his bobbing Adam's apple with great splashes of tea. He leapt to his feet continuously, spearing the biggest pieces of everything on the table before anyone had a chance to pass a plate to him.

He would have been tall if he'd stretched to his full height, but his shoulders bowed over and his almost hairless egg-shaped head, set on a long skinny neck, seemed to grow straight out of his chest cavity. He lollupped along, bobbing head, piston-like arms and long bouncing legs, each anatomical part seeming to move in rhythm to the sound of different drums.

"He looks like a turkey." Mary's succinct remark was lost on Cecil as he busied himself getting settled in the hired man's cabin. His mouth accommodated a splayed set of teeth that would have been comfortable in a mouth twice as large. A front tooth was missing, and through the space he whistled a tuneless noise like a boy afraid of the dark.

"I like to hear him laugh!" Willie was not impressed by appearances, but he was fascinated by the burbling "Yo-ick-a-honk" that went for mirth and issued from Cecil's innards constantly.

Sam, a socialist at heart, considered all men equal. Contrary to the established British custom of that time, he insisted that hired help, family and visitors sit at the same table. He regretted his principles after the first meal with the new man. Besides his atrocious manners, Cecil was quite deaf and punctuated all unrelated table remarks with bursts of his obscene laughter.

Tension grew unbearable. Sally's head bent in disgust, Sam grew apoplectic and Kate, a stickler for proper table manners, tried to shush the children. They all roared with laughter after every "Yo-ick-a-honk" from Cecil.

"Shut up!" shouted Sam. For the length of a minute all was quiet. Then Willie leaned across the table. "Laugh again, Cecil! Laugh again!" he begged.

Sam reached over, lifted his son, smacked his bottom soundly and set him howling back in his place. Cecil subsided, looking forlorn and alone.

As a farm hand he was hopeless.

"Go to the south pasture and bring home the cows," ordered Sam, and away went Cecil over the hill.

Sam rushed in at suppertime. "Where's Cecil?"

A burbling guffaw issued from the hired man's cabin by the kitchen door.

"He's out reading the Buster Brown funny papers," supplied Mary.

"They dawun't want to come," explained the unrepentant reader.

"GET THEM — NOW!" roared Sam.

Cecil scrambled out of the bed. He was off like a loping rabbit, up the hill, only to find the cows with bursting udders swaying down the trail towards him, mooing and making for the water trough.

That fall Sam's reputation for yelling topped his reputation for speed.

"That Marks sounds as if he's going to kill that new man of his," remarked the section foreman in the village in the valley below. "You can hear him yelling for miles."

One morning Sam's patience wore out. He paid Cecil off and dumped him in town. Men were scarce that year, and by the afternoon Cecil was hired as a section hand on the railroad.

First day on the job he nicked a fellow worker's toe with a pick, let the train run over the foreman's axe, and dumped a kit of tools in the river. The second morning he was left to remove the "Jigger," the wheeled vehicle that carried the section men to and from their work, from the track.

The moment the crew moved around the bend to work, Cecil stretched out by the track for a snooze. A few minutes later the warning whistle of an approaching train galvanized him into action.

"Sive me dinner pile! Sive me dinner pile!" he shouted, and grabbing his own lunch pail from the top of the jigger, he leapt for the ditch as the freight bore down on him.

The train hit the jigger with a crash heard a mile away, and the men walked dinnerless five miles back to town.

"I wonder why Sam Marks DIDN't kill him?" remarked the foreman to the other men, after Cecil had been paid and put on a train to Regina.

CHAPTER 6

Hired men were always available, due to the influx of Europeans hungry for land and farm experience. Few resented Sam's yelling. Most wanted to stay on after harvest was over.

Kate, who needed help the most, couldn't get it. Hired girls came and went with monotonous frequency.

"There's a greasy film on your lamp chimneys." Kate held the glass lamp globe up to the window. "Polish them with a linen towel," she instructed Sarah Jane, "after you've wiped out the soot and washed them well."

"Ain't no linen dishtowels I can see," Sarah Jane muttered.

"Where are the new flour sacks?"

"You told me yesterday to put them by for the baby's nappies." Sarah Jane splashed dishwater and looked sulky.

"Oh." Kate hesitated. "Well, never mind then. Go upstairs and make the beds and see the undersheets are smooth or you'll have lumpy beds.

"It's lumpy flour sack sheets as make lumpy beds," retorted Sarah Jane.

Kate threw out the dishwater, got a new supply and set vigorously to work re-doing the dishes. Sarah Jane went upstairs and packed.

"Never mind bringing any more hired girls from town, Sam. I'll just dive into the work myself." And dive she did. When the dishes were done properly, the beds got left. When preserving was finished, the washing wasn't done. She mended and sewed and the weeds choked the plants out in the garden.

She wrote to her sister in the east. "I'm desperate for help, Jessie, Mother promised she'd come next year. But would you come for a few months now if we send you your fare? I'm sure you'd like it like we do."

Jessie arrived later with one small suitcase of clothing and one large trunkful of reading material. Kate was delighted. She revelled in the stories, the styles, and the recipes — especially the recipes.

"Just listen to this, Jessie. Doesn't it sound delicious?" She read aloud one after another. They all required things seldom found among the simple ingredients in a prairie pioneer cupboard. But no recipe fazed her. She was a born innovator. She substituted saskatoons for apples, wild gooseberries for cherries, and rhubarb (which grew everywhere) for everything.

She didn't hold much for measurements either. A glance at the instructions and she waded in. If it called for a cup of sugar, two wouldn't go amiss; if sugar was scarce, a half was dumped in. If it called for a pinch, what was wrong with a handful, and if the baby cried for attention, something got left out altogether.

Her experiments always lacked perfection, but she was never discouraged by failure. The mistakes were tried on the kids, the cats, the dogs — in that order. If all else failed, Sam's pigs got the lot.

"That recipe is no good! I've a good mind to write to that magazine and give them a piece of my mind. Why don't they test their recipes before they print them!" Satisfied that the blame was put in the proper place, she would close the magazine firmly and put it away for the next experiment. Her peach cordial was most remembered.

"Jessie! I think I'll try this cordial!" The children were napping, the house was immaculate and the two sisters sat in the sunny front room overlooking the Qu'Appelle valley.

"Mmmm." Jessie noted the picture that went with the recipe — several fashionably dressed women sipping at tall glasses around a dainty tea table. Kate, magazine in hand, pulled on a fresh apron.

"I'd like to have something special to serve when Mother comes out. Have some of the neighbours in to meet her. This peach cordial would be just the thing for an afternoon tea." She set to work at once, stoking the stove and pulling out pans and bottles.

"But Kate, you've no peaches."

"I've got chokecherries, left from last year. I want to get them used up anyway. They are all fruit — same thing — I could add some other things, too. I've got a few jars left, I'm sure," she enthused. Jessie looked skeptical.

"You put it in crocks, first, it says here." Kate glanced at the list of ingredients. "Let's see. I'll need the butter crock and a few bottles — some sugar too — mmm —." Happy as a bird on its first flight, she sloshed in chokecherries, wild cranberry preserves, a little corn syrup to make the sugar go further, and lastly, some raisins. "Raisins always give it body," she explained.

Later, flushed from climbing up and down the cellar steps, she added: "It must stand a few days. Then you draw off the liquid, bottle it and put it in a cool place."

Her timing was as elastic as her measurements. The cordial was forgotten in the busy weeks that flew by. Jessie went off to visit a neighbour and the housework piled up again. On her return, the sisters cleaned house together and prepared to do a mammoth wash.

The day was sweltering. Kate propped windows and doors open to catch any breeze. She stoked the old cookstove with firewood till the iron top glowed bright red. Jessie carried in pails of rainwater from the barrel outside by the corner of the house, where it caught the drip when it rained. She put the water on to heat in pots on the stove top. A copper boiler, well-laced with homemade lye soap, bubbled and boiled over soiled white tablecloths and towels.

They dragged in the oblong wooden washing tub and filled it with hot water. It sat on four high legs and had a corrugated circular bottom. A slatted wooden rocker, also with a circular bottom, had a high basket-type handle that had to be hand-propelled. Kate fitted it down into the water and tossed in armful after armfull of dirty clothes.

"Drat the luck! This thing is stuck again! Why don't these devilish inventions work smoothly?" Buttons spanged out and materials jammed between the roller each time the rocker moved.

Jessie with patience helped to unsnarl the clothes.

"Anyway, it's better than doing a washing by hand. Here, let me take a turn." Jessie moved the rocker back and forth while Kate took a rest and fanned herself.

Jessie paused. "What's that nice fruit smell? I've been noticing it all morning."

"Heavens! My cordial! I forgot all about it." Kate dried her hands on her apron. "Let's try it and see if it's ready."

They lifted the first crock, skimmed and ladled the liquid into two glasses and sat down to try it. Jessie suspected it was somewhat stronger than it should be.

"My, that's good!" exclaimed Kate. "I knew it would work."

"It's worked, all right," said Jessie, and the two sisters finished the glasses and went on with the washing.

"Let's try some more," Kate suggested later. By the time the washing was hanging on the line, she was singing old Scottish ballads at the top of her lungs.

Nearing suppertime she said to her sister, "You know, Jeshie, I think I'll lie down for a bit. I've a bit of a headache. I shernly feel shleepy." That was the last anyone saw of her for the rest of the day.

As the summer sun topped the prairies' rim, Kate crept out of bed and tip-toed down the cellar steps. She was very busy for the next while. When Sam came down to light the fire in the kitchen stove, breakfast was already on the table.

On the way out to feed the stock, he stopped in amazement to have a closer look at his prize pigs. They staggered about, bumping into each other, looking cross-eyed or trying to stand on three legs. They kept toppling forward on their snouts with contented grunts.

Kate, like her Scottish forebears, couldn't abide waste. She had dumped the cordial in the pig trough. Sam wisely refrained from mentioning it. The pigs showed no ill effects from their indulgence, but Kate went to bed that night with another splitting headache.

CHAPTER 7

That fall was one of abundant increase. Wheat went an average forty bushels to the acre; the gardens were bursting with produce including a bumper crop of cabbages, and Kate was pregnant again. Jessie had returned to the east and Kate was lonely.

"I think I'll call on that new Schneider family next door this afternoon, Sam. I'll take over some of these cabbages. Welcome them to the district." She bagged the vegetables and placed them in the bottom of the baby carriage. The baby was rolled in a shawl and placed on top. With two toddlers at her side, she trudged across the prairie towards her new neighbours' sod shack.

"Velcome! Velcome!" A tall, applecheeked woman enveloped in a snowy apron opened the door. Rich smells of cooking wafted out. Inside, the uneven walls glistened with a fresh coat of whitewash. Everything smelt of soap and water cleanliness.

"Ach, Missis! You sit!" Mrs. Schneider cast anxious looks at Kate's pale face. "I make tea. Some streudel, too." She placed a cushion at Kate's back, put the baby on a clean rag rug on the floor and herded the two children outdoors to join two of her own in the yard.

"Goot cabbage, Goot! I like! You make sauerkraut, Missis? I go tomorrow to you and show how. Good for body!" She laughed good-naturedly as she filled steaming cups and sliced slabs of cake.

The cake had strange new flavours which Kate appreciated.

"Very good," She helped herself to a second slice. "Might I have the recipe?"

"You like, Missis? You write!" The jolly woman, pleased to be appreciated, brought paper and pencil to copy the recipe.

Loud yelps outside interrupted their quiet conversation. A thump at the door announced another visitor. A fat woman, carrot-red hair straggling out from under a man's peaked cap, stuck her head in. A man's suit jacket was held together over her ample bosom by a large safety pin. She stumbled into the room. Laces dangled from her too large boots and tripped her at every step.

"Larch is the name," supplied the newcomer. She wiped her nose on the sleeve of her jacket. "Y'ere's me son, Jarge." She yanked a wizened two-year-old out from behind her. He howled and tried to hide in his mother's grey wrapper, which trailed the ground and was stiff with dirt.

"Jarge! Jarge! Ye little booger! Out wi' ye or I'll thump ye oncet I git ye home!" She gave the unfortunate young one a smack on the side of the head that knocked him onto the floor. With a couple of whimpers he was back again burrowing in his mother's skirts.

The woman flopped onto a chair by Kate while tea was being poured. "Missus Marks, be ye? Called on ye jist now, to borry a cuppa sugar." She leaned over and exposed her toothless gums in a wide smile. "Hear ye be a nurse of sorts. I'll be needin' ye one o' these nights soon." She patted her rotund belly and bared her gums again. "Can't pay ye much. Pat says ye don't need it anyway — ye got lots of brass, Pat says." She laughed at her own remarks and, wolfing some more cake, proffered a tin cup for the loan of the sugar.

"Come on, Jarge, ye little booger, or I'll skelp ye baldheaded." With a few jerks of the little fellow's arm, she was out the door and waddling across the prairie.

"Lots of brass," sniffed Kate to herself. "She's got the brass all right. Thinking I'm a public servant." The day had lost some of its sparkle as she made her way home.

"Missus, you make sauerkraut?" Mrs. Schneider, her German neighbour, arrived promptly the next morning, armed with a cabbage grater, a large butcher knife and a stone crock. She sat down, shredded the cabbage and showed Kate how to salt and press it. Two big wooden pickle barrels in the cellar were washed and scalded. Every spare moment between washings, bed making, meals and child care, Kate sliced cabbage until the two barrels were full.

"I don't like sauerkraut," objected Sam.

"It's the only way to use up all these cabbages," reasoned his wife, "and besides, it's good for rheumatism."

"The smell makes me sick, and I don't have rheumatism ——"

It was too late to quit now. She finished her project and the barrels were stored in the cellar to work. Every day she climbed down the cellar steps to take the lids off, squash down the cabbage, and see if her experiment was ripening.

From her motherly neighbour she received other recipes, German dumplings, streudel, goulash and black bread. She flew through her other housework to get started on these experiments.

"Don't throw out that pig's head," she ordered at the next killing. With recipe in hand she launched into the making of headcheese. For hours the kitchen was full of the smell of steamy boiling fat. The sight of skinned white pig heads, with staring disembodied eyes and grinning mouths, littered the cupboards every time pigs were killed.

By some miracle her headcheese was always a success. Not so the sauerkraut experiment. It failed the complete endorsation of the family and was only tried once.

As the days progressed towards winter the thick odour of rotting cabbage grew stronger. Kate, who during her pregnancies turned green at the mere mention of cigar smoke, seemed impervious to the smell that permeated every corner of the house.

Her confinement threatened to be a difficult one. On the doctor's advice and with the uncertainty of the weather, she stayed in town for some weeks before the birth of her fourth child. When she returned home and had time to check, she said, "Where's the sauerkraut, Sam?"

"I couldn't stand the smell any longer, Kate," he apologized, "it made me sick. The German lady next door said it was so good I gave her one barrel. I buried the other in the garden."

"What a waste," said Kate, with no rancour. "Anyway, the cabbage was all used up." Since she had processed it, her conscience was clear. "But you should have put it in the swill barrel. The pigs would have liked it."

The house had a clean, fresh smell, with an added dash of the perfume of baby soap and talcum powder. She picked up Hazel with a smile and held her close.

CHAPTER 8

"Light the lamps, will you, Sam?" Kate rocked and nursed her new-born child as the bleak December night closed down. Outside the wind rose, swirling the newly fallen snow into conical drifts across the yard.

"Time for your bed, children. Eat up your supper now, and off with you." The three children were spooning bread and milk and dreaming out loud of Christmas.

"Look at that blowing snow outside, Sam. I'd hate to be out on a night like this! We'd better put a lamp in the window in case anyone gets lost out there." Kate buttoned her blouse, burped the baby and put her back in the cradle. She looked around her with satisfaction. Sam had finished the chores. The place was clean. The house was snug and warm and the cellar was full.

Her reverie came to an abrupt end with the sudden sound of sleighbells and the screech of runners on sandy snow outside. After a loud thump on the door, a squat, grizzled little man staggered into the kitchen. He peeled chunks of ice from his moustache and spilled snow and slush across the clean kitchen floor. The strong stale smell of whiskey came in with him.

"What seems to be wrong, Larch?" Sam closed the door and directed the children up to their beds.

"Wife's in a bad way, Sam. 'Er time's come. Beggin' fer yer wife to come."

Kate remembered the red-headed woman whom she had met in the fall.

"You must call a doctor, Mr. Larch. I'm not a trained nurse. I've never attended a birth before.

"You be the only nurse around. She be wantin' you."

"My wife has a newborn baby herself. Can't you call on one of the other neighbours." Sam swept up the snow and hurried the children up the stairs.

"None'll come close by."

Kate, heart pounding, grabbed an empty flour sack and began to stuff clean sheets, dusters and rags into it. She rushed from one room to the other, collecting things as she went. Sam followed her into their downstairs bedroom.

"You shouldn't go out in this storm, Kate. What about the baby's feeding?" he whispered.

"What else can I do?" She shoved some of her own children's things into the bag. "She won't likely have any clothes for the baby from the look of her last fall. I'll just throw in a clean blanket and some sheets. I can't leave the poor soul in the lurch."

"I'll dilute some of the cow's milk for the baby until you get back." Sam lifted the heavy sacks and shifted them to the sleigh box outside. Kate tied a scarf around her hat, pulled on her husband's sheepskin coat, some mitts and overshoes, and she was ready. She followed the gnarled little man out into the bitter night and climbed into the sleigh box beside him.

The trip across the frozen fields was made without mishap. Kate sat stiffly, facing the biting cold. Anxiety for what lay ahead knotted her insides. Occasionally Larch muttered disjointed words, pounded his horses, or whistled snatches of Irish songs into the moaning wind.

"You go on into town for the doctor. I'll keep things going until you get back. And hurry!" She shouted the order over her shoulder when the horses stopped in their own yard. She climbed down from the sleigh, and without a backward look floundered through a snowbank towards the little sod shack.

The muggy stench of unwashed bodies, smoking kerosene lamps and raw manure struck her when she opened the door. Faint flickers of light came from a barn lantern and a soot-encrusted lamp sitting on a table in the centre of the room. Mrs. Larch lay on a bed against the wall. Only her tangled mop of red hair showed among the rags and coats heaped on top of her. The moment Kate appeared the woman let out a few screams, well larded with curses.

Kate marched over to the bed.

"Be quiet, she commanded. "If you don't stop that swearing I'll go home at once! AND I MEAN IT!"

There was silence for a moment. Then a half-howl, half-whimper came from under the bed. The little boy crawled out, naked to the waist and covered with dirt and excrement.

"What in heaven's name will I do with this filthy mess? Dear God! What WILL I do?" For a moment she hesitated. Where to begin?"

To one side of the bed was a large cookstove. The chill that pervaded the room told her fire was almost out. She grabbed some blocks of wood and rattled the shaker. In a moment tongues of fresh flame began to crackle under the kettle of water she had filled from a large pail.

She filled a tin basin with warmed water, stripped the child and washed him off. Soon he was clean, dressed in a shirt and napkin from the supply she had brought. He sucked ravenously at a bottle of warm milk she fixed for him, and was soon sound asleep in a box near the stove.

She heated more water and faced the job of bathing the woman. After the bed was stripped to the bare straw tick and the bedclothes and rags were piled in a corner, she set to work. With some help from Mrs. Larch, between labour spasms, the job was done. Putting her own clean sheets on the bed, Kate covered the woman with the clean blanket and on top put a horse-blanket which seemed less filthy than anything else in the room.

A strong, persistent, barnyard stench puzzled her. She lifted the rags that had been on the bed and some flat, mushy, cloth-covered clumps rolled onto the floor.

"Hey! Don't throw out me cow mook!" The woman sat up in bed. "Took me all mornin' follerin' the cows around the byre to git it. It's gotta be soft and warm else it's no good." She tried to yank the offending cloth-wrapped clumps from Kate's hands.

"What!" shouted Kate. "What do you use it for?"

"Best cure for birthin' pains there is! I allus puts it on me stummick at the first signs when I'm birthin'. I jist wrap it arount me middle an' it works great." She grinned a gummy smile at Kate and hung on.

The cleanup process was helped by another labour spasm. She let go the offending cloths and grabbed for the bedstead instead.

"Nonsense!" Kate seized her advantage and stuffed some of the clumps into the firebox, hoping they wouldn't put the fire out. "You're not using any of that filthy stuff while I'm here!" The rest, she opened the door and heaved out into the night.

"Whenever will that doctor get here?" Gusts of wind and snow tore at her skirts and slammed the door wide open. The shadowy shape of the sleigh loomed up in the darkness. It was exactly where she had stepped from it earlier. There was no sign of Larch or the horses.

"Doctor won't be comin', not likely." Mrs. Larch sat up in bed. "Larch 'ud never pay 'im before. 'E's proud, is Larch. Wouldn't go fer 'im fer fear of bein' refused."

"Well, where in heaven's name is your husband now?"

"Oh, 'e's out in the cowbyre with a drop o' whiskey, like as not. Larch never could stand the sight o'me in labour. 'E's sensitive, Larch is. 'Ard on 'im. Allus 'as to 'ave 'is whiskey at these times. Allus before I was mostly alone. Larch never were much help. I've 'ad five o' them an' oney Jarge 'ere livin'." She wiped her eyes and nose on the edge of the clean sheet. "There won't be no doctor now."

"No doctor! NO DOCTOR!" the gravity of her position struck Kate anew. She was alone with no hope of help, and from all appearances the birth would be a hard one.

"The nerve of that man ——" Her words were drowned out by a muffled scream from her patient.

"Just hang onto my hands now. Take a deep breath! There, now. that's better." She wrung out a wet towel and wiped the perspiration from the face and hands of the agonized woman.

The clock ticked on. It was well after midnight. It was useless going out in the storm to try to alert the husband. If she succeeded in reaching the barn he would probably be too drunk to be of any help anyway. She tried to remember what had been done for herself.

"Here's some coffee and some hot milk." Mrs. Larch sat up and drank greedily.

"Larch was allus after me to use a quill. 'Use a quill, Moll', he'd say. 'Use a quill.' And oncet he did use it on me but it don't work good fer me. His fambly allus used it, Larch says. Worked fer them, too, but it don't fer me." She took another swallow of coffee and grinned.

Kate grabbed the half empty cup as another pain began.

"What do you mean by a quill?" Mrs. Larch was sitting up again and ready for more coffee.

"Oh Larch 'ud get a cock's quill and rub it good with pepper and when I'd get a pain 'e'd 'old it under me nose. It's supposed to make yuh sneeze and the baby pops out easy-like. Never did work fer me, though. Larch allus swears by it, 'e does." She shook her head in a puzzled way.

The long night hours crept slowly by. Kate, exhausted from worry and the need to nurse her own child, sank into a chair by the bed. Her eyes closed and she was instantly asleep, only to be awakened by the muffled sounds of agony by the woman in labour.

The wind died. Night slowly melted into morning. Rosy light glittered on the ice-rimmed window and lit up the ugly corners of this dismal room. Kate replenished the fire, boiled more water, and prayed.

"If someone would only come!" She sank onto a chair, half-sobbing — too tired to comfort the woman at the next spasm. An extra loud scream galvanized her into action.

"Hang on! Breathe through your mouth! Breathe through your mouth! That's good now. Hang on tight!" The next few minutes were desperate and required all her attention. A baby girl announced its arrival with a resonant howl. Kate, no longer weary, stared down at the tiny mite.

"I did it! I did it myself. I delivered a baby!" With trembling hands she cut the umbilical cord with the scissors she had boiled for the purpose.

"You have a fine baby girl, Mrs. Larch." She wrapped the child in a towel and held it down so the mother could see.

"Don't fergit the salt pork! It's in the cupboard with the grub." Mrs. Larch, her eyes swollen shut, tried to sit up.

"WHAT?" Kate stopped her task of the moment. "What did you say?" Was the woman delirious? If someone would only come, she could send for the doctor.

"The salt pork! Git the salt pork." The mother tried to lift her head from the pillow, with little success. "The salt pork," she insisted.

"Whatever for?" Kate whisked the baby through the bath and laid it in a basket she had prepared on the table. "Better to humour her for now," she thought.

"Fer the cord! Yuh cut the cord. Yuh gotta tie salt pork to the cord soon's it's cut. Do it afore it's too late."

"There! There now. You just lie still and I'll fix you up comfortably until we can get the doctor."

"Yuh gotta git it." Mrs. Larch was almost sobbing. "I ALLUS tie it to the cord! When the cord dries up it's rotted and falls off too. Helps in the dryin' process. Gives the baby more strenth tuh suck. Makes it stronger almost at oncet! It's a old remedy — come down in my fambly — I've did it to all my children. Never fails!" She sank back, exhausted.

"You leave everything to me." Kate poured hot milk into a cup, added sugar and helped the woman to drink. Mrs. Larch swallowed greedily, lay back and was asleep in a moment.

"Salt pork, indeed! No wonder only one survived."

In a few minutes the mother, toothless mouth wide open and still snoring noisily, lay on a cleaned, smoothed bed, her new-born infant asleep by her side. The floor was swept, the table scrubbed and the stove lids wiped clean.

The little boy stirred and sat up in his box bed. He stared at Kate with wide frightened eyes, and began to whimper. She washed and dressed him in a clean woolen suit from her flour sack supplies. He gobbled a bowl of warm porridge, climbed down and hummed happily to himself from a blanket spread on the floor.

The sun shone in through the melting ice on the window. The kettle sang on the back of the stove. Kate, no longer anxious, sat down on a chair and smiled.

At the welcome sound of harness jingling and runners squeaking on the snow, she rushed to open the door.

"You all right, Kate?" Sam handed the lines over to Mr. Schneider who was with him, and leapt out of the sleigh. "You all right? Where's Larch?"

"I guess I'm all right. I'm pretty tired." Kate's voice shook. Now that help was at hand she must not break down.

"Larch is in the barn, I think — but we have a new baby girl here ——"

She briefly explained the night's ordeal. "But this woman must get to a hospital. She was delirious — I think — before she fell asleep."

"No use to rouse that jackass of a husband. Kate, you get your things on and we'll make a stretcher and put her and the baby in the sleigh. Mrs. Schneider is looking after our children until we get back."

"My Anna will go. You come home, Missis." Schneider helped the transfer of the sick woman to the sleigh box, scarcely waking her in the process. The little boy was taken along with them.

Home once more, Sam helped his weary wife into the kitchen. Mrs. Schneider put on her winter wraps hurriedly.

"You stay here, Mr. Marks. Stay with wife. I go with husband to hospital. Foolish Larch. I vould haf come. Missis with a new baby herself. Missis, you rest. Very bad night for you. I take boy with me."

Sam fired up the stove, made tea and buttered toast. Kate undressed and nursed her baby in bed. She recounted the night's trial. Sam listened angrily. "If this ever happens again I'll put my foot down. The gall of that jackass. I thought the fool had called the doctor already. Never dreamed he'd leave it all to you. He owes Schneider for seed wheat he got a year ago. Won't pay for anything. Guess that's why they didn't call on them to go."

Kate put the baby in the cradle and sank back exhausted on the pillows. "Now that I'm home I can see the funny side of it! Those old superstitions! It's the salt pork that finished me." Kate laughed hysterically. "Now I don't know if she was delirious or if it was just another home cure!"

Later Sam poured another cup of tea for his sleepy wife and brought it in. "I'm sorry you had to go through this, Kate. It'll never happen again."

"Nonsense!" Kate sat up with a jerk. "I wouldn't have missed it for the world. Besides, I couldn't leave that poor soul in the lurch. Who WOULD go to her help. I'm going right to the hospital tomorrow and find out about some things. Next time I'll know exactly what to do!"

One morning two weeks later a sleigh drew up to the door and out climbed Mrs. Larch, the baby in her arms. Her face was clean and her red hair was smooth and tidy beneath her woolen hat.

"You wait there, Larch! I'll be out in a minute," she called to her husband.

Kate opened the door and helped her to a chair.

"When did you leave the hospital?"

"They 'lowed me out today. 'Ad to get 'ome to Larch. 'E an' the boy needs me. Larch Misses me, 'e does. Works 'is fingers to the bone for us, does Larch. Just thought I'd show yuh 'ow we wuz before we leave fer out west, tomorra. Show yuh my Anna, too."

"What a nice name." Kate looked at the wee face and the clean blanket and wondered how long it would take to get filthy.

"Called it Anna after Missis Schneider. Wouldn't 'a made it if it 'adn't been fer her. Took me all the way into the 'ospital, she did. Saved me life, Larch says. 'E's ever so grateful to 'er, 'e is. I'll say that fer Larch. 'E never fergits a kindness, does Larch. 'Call 'er after Mrs. Schneider, Moll! Sort of a reward fer 'er kindness,' 'e says. No, Sir! Larch never fergits a kindness." She rewrapped the baby and waddled back out to the sleigh.

"Well, well! I never!" As the sound of the sleigh bells receded, Kate burst out laughing. "Well, I never!" she howled. "Wait until Sam hears this!"

CHAPTER 9

Kate opened the winter mail.

"Mother is coming in the spring, Sam!" She was jubilant. At last her reason for coming west would be justified.

A small house close by was hauled in. Every extra minute weaned from

baby-caring, housework and farm duties, Kate used for scouring, papering and painting the place where her mother would live. In the spring she planted nasturtiums and sweet peas around the front. Bright calendar pictures adorned the walls and floursack curtains were hung at the windows.

"I wonder if she'll like it?" she fretted. "I'll have some of the neighbours in for tea to meet her. I hope she will find the Schneiders and Gordons as interesting as I do." She surveyed her work worriedly.

"She'll be sure of three square meals a day, whether the neighbours are tony enough for her or not."

Sam's concern for the basics took precedence over style. He was annoyed and vaguely uncomfortable when his mother-in-law's background was mentioned. She had been named Margaret Isobel and was the only daughter of one of the three Fraeser sisters who claimed to be direct descendants of Simon Fraeser, Lord Lovat. Her family always made much of the fact.

Sam heard the story often. Fraeser had been a huge Falstaffian wit, full of deviltry and daring. These attributes he used to confound and upset the hated English at the time of Bonnie Prince Charlie, whose doomed cause he had championed. He also won historical distinction by being the last man in the British Isles to have his head chopped off as a means of shutting his mouth, something this witty old nobleman had difficulty in doing. He was led to the block amid a great crowd of onlookers.

"I've had good use of this head for some eighty years, and now you are welcome to it!" Fraeser announced to the axeman, as he took his last pinch of snuff. His wide, slanted, green eyes were bright with mischief, and undimmed with tears. A broad smile lit up his puckish face as he stepped forward to his death. The chopping block, last stained by his royal blood, now dark with age, sits near the crown jewels in the Tower of London.

Kate's mother, orphaned early in life, was adopted by her aunts, the two remaining Fraeser sisters. She resented their stiff, unbending discipline, and in the face of great opposition married Hugh McDougall, a gregarious Highlander, trained in stonemasonry. Their first child, Catherine (Katie) Cameron McDougall, was born in the fall of 1882. The following spring the family sailed for Canada, and settled in southern Ontario.

"Father was successful in Hamilton, but when they moved to Bala it was a disaster," Kate told her husband. "It was fine in the summer, and Father helped to build many public buildings in the area. He built Mother a lovely house, Cromarty Lodge they called it, overlooking the water. We all worked for the summer visitors who came up from Toronto."

"During the winter, jobs and money were both scarce. Father and some of the men clubbed together and bought an expensive saw for ice-cutting. The ice was cut into blocks and stored in sawdust for use in the summer. One day in February when they were working on the lake the saw slipped into a hole in the ice and sank to the bottom. Father, without hesitation, dived into the freezing water and brought the saw to the surface. The men hauled him out and carried him shoulder high to the local bar, where they proceeded to drink to his health. He sat there in frozen clothes and made good progress towards losing it — he developed pneumonia and died shortly after. I've never had any use for liquor since then, Sam. I'm so glad you've given it up for good!"

"I was only seventeen when he died, and I kept the family until I came out west. The boys were working by then. I'm so glad Mother is coming here. I'm sure it will be better for her health."

On her mother's arrival date, Kate surveyed her handiwork. She was anxious for her mother's approval. "It's as clean as soap and water can make it, anyway. I do hope she'll like it!"

Her satisfaction with the house didn't apply to her own appearance. She squinted in the bedroom mirror, undid the heavy coil of braided hair from the top of her head, and scraped a comb through it. It had been a bright russet colour that had added to her attraction when she had arrived from the east. It was now faded to a dull mouse colour from the dry climate and her own neglect.

Her eyesight seemed to be getting worse. She had great difficulty in seeing the film of dust which constantly gathered on everything, but her failing sight didn't hide from her the furrowed brow and the frown lines deepening between her myopic eyes. Her mouth, too often of late, was set in a hard, straight line.

Her mother, ignorant of modern child psychology, had always been blunt to her children about their shortcomings.

"Too bad you're so plain, Kate. Jessie got all the looks in the family." These oft-repeated words echoed hopelessly in the back of Kate's mind now. She thought of her father's rough, homely features as she gazed at her own.

"Mother always said I inherited the McDougall looks and the Fraeser temperament," she said.

"Mmmm!" murmured Sam, through a screen of shaving lather. He stood by the bedroom washstand, slapping his straight-edged razor back and forth on the leather strap.

"You look all right to me." He carefully scraped down his cheeks, and around the black moutache which covered his upper lip. He wiped the razor clean on chunks of newspaper, and gave Kate a friendly hug.

"What's worrying you?" He rubbed his freshly shaven face against her cheek and smiled over her shoulder into the mirror. She smiled back and felt better.

"Oh, nothing — I guess." She smoothed the long folds of her blue taffeta dress. It had been her best for several years now, between pregnancies. It swept over her well-corsetted figure to her ankles. She added a white lace collar, and put on a large yellow leghorn hat. It was turned up at the side in the style of a South African army hat, and was held in place by a huge blue ostrich plume.

Sam went to get the team. Kate, smiling confidently, prepared the children for the trip to town to meet their grandmother for the first time.

"The best laid plans of mice and men gang aft a-gley" — Burns' immortal words — covered the arrival of her mother.

On the way to town, little Ruth threw up on Kate's dress; Willie slapped Mary and made her cry, and the baby woke up and howled all the rest of the way. Kate's yellow hat blew off, frightened the team, and was run over by the wheels of the democrat. As the train drew into the station, her clothing looked rumpled, Sam's patience was wearing at the edges, and the children's faces were dirty from an avalanche of tears well mixed with prairie road dust.

Jessie stepped from the train first. With the conductor's aid, she helped down her tiny, regal mother, who stood, straight as a pine tree, encased in a swath of rustling black taffeta. A black widow's bonnet, edged in fluted silk and tied under the chin, framed a face cool, serene and lovely as a winter's lake. From under its brim, a pair of smokey grey eyes stared out at the untidy lineup of her daughter's family waiting to greet her.

The moment her foot touched the platform, the children, who had tumbled forward to welcome her, drew back, astonished and in awe of this beauty who was their new grandmother.

The trip home bordered on disaster. The two older children had to be squeezed into the back beside the tired travellers, and later put at the back to swing their legs over the open end of the democrat. Sam angrily drove the horses at their old livery stable speed, and Kate fretted all the way home for fear the children might fall off on the road.

At home, Kate opened the door of the tiny house she had spent such labour fixing, and was appalled at how bare and ugly it now looked, as she mentally compared it to the pretty home they had had in Bala overlooking the lake and wooded hills.

Her mother walked in, surveyed her new home, then wordlessly proceeded to remove her bonnet and shawl.

Kate was grateful for the books and magazines brought from Ontario. Her mother, in the early stages of tuberculosis, suffered from the harsh prairie winds. In the spring they prepared to leave for British Columbia, where Jessie had secured a position in a post office.

On the morning of their departure Sam drove them to the station. Kate got out and walked to the train, hoping for some word of encouragement. Her mother turned on the steps of the platform and remarked, "You have quite enought to do now, Kate. See that you have no more children."

"She's a bit late," wept Kate on the way home. "We are having another one, sometime in the early winter, I think."

CHAPTER 10

The crops were good that fall. Two new granaries were built. After some hesitation, Sam added a kitchen to the house. Above the kitchen was a new bedroom. The added space provided a spare room for winter visitors. It was common practice that during the winter families sometimes came to stay for several days. It broke the monotony of months of bitter cold. Kate had not had this pleasure because of the cramped space. She was now delighted, and pored over the Eaton's catalogue, more interested in the spare room equipment than the expected baby's layette.

She ordered a new washstand and a toilet set containing a basin, jug, soap-dish and lidded chamber pot. Each piece of crockery was decorated with large blue poppies. When they came, they were put in the new bedroom at once.

Polly, a quiet, efficient, Finnish girl, came to help with the housework during Kate's confinement.

"You can sleep in the spare room, Polly, until we have visitors," Kate decided. The children were fascinated by the new bedroom and all it contained.

"Aren't they beautiful?" Mary clanged the pot lid and tested out the tiny washstand door.

"I like the nice soap Polly uses. Mmm, doesn't it smell lovely?" Ruth pressed the perfumed cake to her cheek.

"Come out of Polly's room at once," scolded Kate from below. Then she turned to Polly.

"We'll just dye some of these new flour bags for the spare room curtains. I've lots of blue dye on hand."

The bags flapped out the flour on the line while Kate dumped pails of water into the copper boiler on top of the stove, and stoked up the fire. With a quick look at directions, she dropped in a package of dye and tested for colour.

"Mmmm — a bit light. We'll put in this dark blue to 'set it', and she threw another packet of dye into the boiling water. The bags were collected from the line and tossed into the water to boil. The job was never completed. A baby boy, christened Hugh, and later nicknamed Buster, announced his intended arrival.

"Just soak them in lye, Polly," said Kate wearily, cradling her newborn son hours later. The streaked, sodden mess had been pried off the boiler bottom. "I'll give that storekeeper a piece of my mind about that dye. Sending stuff like that out, and us having to pay good money for it."

The three older children crowded into the bedroom to see their new brother.

"He looks like a baby puppy, doesn't he?" Willie looked down into the wrinkled, squirming face.

"Yes, but he sounds more like a thirsty cow." Ruth stood by, fascinated. Buster let out his howl of protest to the world and established the family record for resonant bellowing.

Polly was busy the next few days caring for Kate and the baby, baking, cleaning and keeping the curious children out of the spare bedroom. Her love life and the equipment in the spare room shared equal interest.

"I think Polly should marry Mr. Carney. He has such lovely eyes." Mary nursed her doll and looked dreamy.

"I like John the best. His hands are always nice and clean, especially after milking." Ruth was John's favourite.

Sunday evening arrived and with it came John, the main male contender for Polly's favours. Her reputation as a cook travelled like a prairie fire, reaching across the lonely miles into all the bachelor's cabins. Each Sunday evening a growing number of young farmers, scrubbed and hopeful, waited in the parlour for Polly's approval.

The baby had been tucked to sleep beside Kate, and the children whisked off to bed early, before Polly could make her appearance at the parlour door. John had barely removed his hat when two other young men rode into the yard, put their horses in the barn, and joined the suitors' circle. Polly flew in and out, about her household tasks swiftly and silently. Not much given to talk, the long, embarrassed silences didn't bother her. Her apple-red cheeks and an extra brightness in her eyes were the only indications that the situation was unusual and pleasing.

Murmuring voices above floated down the stovepipe hole and indicated an audience. The two little girls were interested in the proceedings and couldn't go to sleep. There was a dead silence in the room below.

"To bed with you," said Polly, slipping up the stairs.

"I'd like a drink of water, please, Polly."

"I'd like one, too."

"Off to sleep with you now," and Polly, her mind on her callers below, held the cups for the two children.

She rejoined the threesome and for some minutes only the ticking of the clock and an embarrassed clearing of dry throats broke the silence.

Suddenly, from above, the clatter of bedroom crockery floated down the stovepipe hole.

The four young people, familiar with the natural equipment of every farm home, sat in scarlet embarrassment at the sound of trickling water and the clang of the pot lid above.

A tiny figure suddenly appeared at the parlour door. "Polly, I hope you don't mind. We used your new chamber pot with the poppies on it. Ours is full."

"I'm going to get up for a while today, I think," said Kate cheerfully the next morning, "as soon as the baby is fed." But it was three more days before she took the first shaky steps.

Sam came into the room shortly after the baby's feeding time. He sat on the edge of the bed and put his arm around his wife.

"Kate, I've bad news for you. Your Mother died last night."

CHAPTER 11

"Willie! What ARE you doing with that syrup?" Kate grabbed the jug as the sticky fluid began to run over Willie's meat and potatoes. Kate was pale and tired. Polly had married her beau, and few hired girls wanted a job in a house with five small children.

"Joe eats syrup on his potatoes," reasoned Willie. "He says it all goes down the same way anyway."

Joe was a wizened little bachelor who lived just over the hill, and who was a constant wonderment to the children. Joe, as all prairie bachelors did when they couldn't stand their own company any longer, visited his neighbours, and wasn't as popular with the adults as with the children.

His farming was a haphazard affair. He "got up when he liked and went to bed when he pleased," and was careless about returning borrowed time or tools.

"You're spending too much time at Joe's," scolded Kate, "You musn't bother him."

"We don't bother him." Willie looked calculatingly around the room.

"He's got everything so handy at his place."

Joe's narrow bunk bed, cook stove, small table and cupboard above it, occupied one corner of his one-room shack. The rest of the room was used for everything — a place to store sacks of grain, or fix sets of harness, or skin pigs after butchering. He once backed a horse up to the door and sat on his bed to do a shoeing job. Hens wandered at will through the open door in summer, pecking grain or crumbs from the floor, or hopping up and helping themselves from the open pots and jam cans on the table.

"Hurry and finish your dinner and get that woodbox filled."

"Joe's got a good way so that he doesn't have to fill a woodbox. He keeps long poles handy and shoves them in the stove as they burn. Why don't you do that, Mama? And he learns us things, too — like this afternoon ——"

"Teaches us," Kate corrected, going on with her work.

Mary kicked Willie under the table. Joe chewed tobacco, of which habit Kate disapproved. That very afternoon Joe had shown the children how useful a good chewer could be, and demonstrated how to stop a hen from jumping into the cream pan left on the floor. He spat a wad of tobacco that arched through the air and caught the unsuspecting fowl a great splat on the side of the head, knocking her off her feet. The children watched, fascinated, as the hen ran screeching out the door.

"Never mind Joe." Kate was exasperated. As the children finished eating and the men were about to sit down to supper, the subject of conversation, singing lustily, drove into the yard and, as was usual, stayed for supper.

"Joe's always glad to get home from town. He never sings on the way in, but he always sings on the way home." Willie had just finished his supper, and he got down to make room for the grownups.

"He always smells like vanilla, too." Mary looked thoughtful. "I like that better than when he smells of 'chew'."

Kate banged down another plate on the table and hurriedly fried more potatoes and cut more bread. The baby began to cry with drum-thumping resonance.

"He doesn't want anything, Mama. He just likes noise. Look! He doesn't even make any tears." Mary picked him up. "Look, he understands talking, too! He's smiling already!" But Kate, weary and impatient, didn't seem to hear.

"You'd better have a visit with Mother and Father for a day or two." Sam helped Kate tuck the children in later that night. "The folks will enjoy seeing the kids. We'll go the day after tomorrow."

"Good idea." Kate brightened up with thoughts of the change. "I'll put the washing to soak overnight, and do it first thing in the morning." She paused. "Sam, you'd better check that stove. I think I smell smoke."

"Everything's fine now, Kate." Sam made another trip to the kitchen. "Now get to sleep and the trip will do you good."

Kate flew around the next morning packing, washing and mending. She was happier than she had been for some time. The children were fed and playing quietly, and the washing tub rocker was splashing the clothes merrily, when she looked out the window. Sam, ashenfaced, was racing towards the house.

"Kate, come in here!" he gasped. "Something terrible has happened!" He closed the door so the children couldn't hear.

"Joe ——"

"That man again." Kate's spirits sagged. "Now what? Something to spoil our trip, I'll be bound."

"His shack burned to the ground — nothing left but a few pieces of tin — and bones — a big patch of grease — that's all ——" Sam retched into the washbasin.

"Don't say anything. Willie will be upset. I'll get Johnnie Gordon to come over."

A little time later the two men started soberly down the road.

"The section men will be coming by on the track in a minute or two. We'll take them over and see what to do."

They waited until the jigger carrying the section men came by, then all six men walked down towards the burned out tragedy that had been the finish of their odd little neighbour.

Tangled wire, bits of glass, and the burned shell of the stove were strewn about the edge of the square where the shack had been. The heap of charred bones drew all eyes. The men removed their caps and for a moment nothing was said.

Johnnie Gordon took command. "We should call the doctor — no, the police maybe. Sam, will you ——"

At that moment a burst of loud profanities came from behind the barn.

Out staggered the unkempt, sodden figure of little Joe. For a moment every man held his breath.

"Where were you, Joe?" Johnnie was the first to speak. "We looked for you in the barn. How did it happen?"

"Dang place caught on fire from the pole I had in the stove. I jist moved out to the strawstack behind the barn there. Had a pig hanging in there and forgot to take 'er out when the fire started. Burned up all my winter's meat." He had rescued a partial bottle of rum, which was still in his hands.

The men all burst out laughing. Joe stood in puzzlement, surveyed his losses, and philosophically took another swig from the almost empty bottle.

CHAPTER 12

"Everybody out of bed. Be quick now, if you want to go to see Grandma and Grandpa."

The first gold-red rays of a late autumn sun hadn't dried the steam on the log house windows.

"We've an all day trip ahead of us. Hurry or we won't get there until the middle of the night."

Kate scurried from kitchen to pantry to cellar, slicing loaves of homemade bread and cold beef, and packing a straw clothes hamper with food for the journey. She put in quart sealers of stewed saskatoons and fresh wild cranberry jam, all picked and canned since the birth of the baby. Bottled milk, still warm from the cow, was put into the basket.

She roused the sleeping children and helped them out of their warm beds, opening dresser drawers and hauling out their "Sunday" clothes as she went. Willie's new brown pants, buckled at the knees, showed two big rips in the seat which had to be mended at once, if only for propriety's sake.

"Drat the luck! How does that child get his clothes torn so often?" She searched frantically for the mending box. "Are they all getting worse at ripping their clothes lately, or is it my fault?" she muttered.

She carried pails of water from the rain barrel outside and filled pots and pans on top of the stove to warm.

"Take down the bathtub, Mary, and Ruth, you get some towels and the baby's toilet soap from my washstand." She ladled out plates of steaming grey oat porridge for the four children. The long tin tub was taken from the back porch wall and set on the kitchen floor, in front of the open oven door.

"It's not Saturday, and I'm not even dirty!" Willie objected, as he was quickly shoved towards immersion.

"Why can't you just bathe the girls? The hired men don't bathe, only when it's spring and there's water in the sloughs."

"Never you mind the hired men's habits, young man." Kate stripped her six-year-old son with little regard for his male pride and had him soaped and washed and wrapped in a towel in a few moments. Quickly the girls had their turns. Soapsuds flew, with small attention paid to stinging eyes or silted corners in ears and necks. One after another, the children gobbled up their porridge, then rushed back to the job of getting ready.

Mary and Ruth brushed and braided each other's hair, put on flounced dresses and clean white pinafores. Willie yanked on a peaked, visored cap over a still wet head, pulled on his recently mended kneepants, buttoned up the suit coat to match, and buckled the belt that was threaded through slots around his lower middle. He smiled at the fine appearance he made in the kitchen mirror, then stooped to brush the dust from his boots. The memory of the smarting indignities of the bath evoked a glower, but it was a weak effort.

Sam shaved, dressed and went out to help milk and feed the stock with Alfie, the hired man left over from the exodus of the harvest crew. The team was hitched to the democrat, and they pranced up to the dooryard. Sheaves of oats and a generous pail of chop (crushed oats and barley) stuck out at the back of the rig. Straw trunks and the food hamper were stuffed into the space between the seats.

Kate swept out, her blue taffeta dress a little too snug and pulling across the front. She pinned on her yellow leghorn hat, crushed from the misuse it had suffered, its blue plume broken in the middle.

"Take your winter coats, girls. It might turn cold before we get back," and she handed the baby and the next smallest up over the wheel to Sam and climbed up after him. The other children clambered into the back, and everyone squeezed down into place.

Sam touched the backs of the horses with the tip of his buggy whip. They were off, spinning down the prairie trail in the cold harvest dawn, stones and dust flying out behind.

The drive was long and tiring. Buster bellowed constantly. The other children, weary from inactivity, fought with each other or howled. Finally over the last few miles they became quiet. The only sounds were the steady thump of the tired horses' hooves, the velvety crunch of the wheels in the trail dust and the gentle clapping of the aspen leaves. The democrat wound in and out through the groves of yellow-red trees. Over a hill and around a curve, they arrived at last at the wide, one-story log house set against the darkening trees. Tiny pinpricks of light were beginning to twinkle at the windows.

In the gloom of approaching night Grandma Marks' tall, well-rounded figure, enveloped in a huge white apron, stood on the squared stone step. The pale lamplight from the kitchen door behind limned her outline in a soft yellow glow. Her silky black hair was piled in a knot on top of her head and topped with a high shell comb.

The rig came to a halt. The children poured down upon her and she held out her arms to receive them.

"Oh, the dear of it! The dear of it!" Black eyes sparkling, she enfolded each tired body in turn.

Her husband rushed out from behind and grabbed the lines from Sam's hands.

"Will! Will! Look sharp, will 'ee? Can't 'ee see 'ee needs be 'eer?" He held the lines in one hand and waved the other one about. The horses moved nervously back and forth, making the unloading process difficult. His son Will sauntered out to join in the noisy mix-up.

Once inside the house, the children stared about them silently. A string of silver pipes snaked across the full length of the room, held near the ceiling with lengths of thin wire. A copper disc closed one end of the pipes where the winter stove would soon be attached. Colourful braided rugs covered every corner of the immense room and gave off muffled crackly sounds. This was from the matting of straw, laid down after the fall threshing, which helped to cushion the floor from the freezing cold of approaching winter.

The log walls, hand-chinked with clay, had a rough, textured look. They were a dazzling white from whitewash, a paste mixture made of milk and lime, used in place of paints by early settlers.

"You like little girls, don't you, Grandma?" Ruth squirmed around, her eyes on the pictures that covered the walls, while her Grandma helped remove outside clothing. Proof of Grandma's devotion showed little girls with chickens, little girls with baskets of flowers, little girls with kittens and little girls with other little girls — framed, coloured prints that smiled down from every wall, with one exception. A big, framed picture of battle in progress, with bursts of cannon, red against a black sky, and along line of fighting, dying soldiers, dwarfed one wall.

A wide yellow sideboard with round wooden shelves to hold the lamps, and backed by a mirror that made everything in the room ripple with reflected light, held a monstrous coconut-covered cake and a frosted glass pitcher full of fresh milk. Pretty bisque figurines of shepherdesses, boys playing flutes, angels playing harps and jolly-faced toby jugs ranged about the cake and invited closer inspection. Warm, spicy smells wafted in from the kitchen.

Kate, speaking regretfully over the loss of her mother, unbuttoned coats, straightened hair and pulled up stockings.

"Thee were good to 'er and thee musn't grieve too long, else thee question the Lord's work."

Kate spoke of the late threshing which, with the new baby to care for as well as the other children, had been a nightmare.

"There were twenty-eight harvesters on the gang," she explained. "It rained incessantly and each time, back they came to be fed until the weather changed. It was Sam's threshing outfit so we were responsible for the men. I felt sometimes that the devil himself lurked around every corner just to plague me."

"Be gentler with thyself, Kate," admonished her mother-in-law. "Thee be too hard on thyself. Be less anxious over things. Lay down a bit, will 'ee, and Ellen will bring 'ee a little som'at to drink. After the children are abed I've som'at to show 'ee."

Kate was soon stretched out resting. Ellen, the hired girl, brought in tea and toast. The happy, sleepy sounds of children and grown-ups, pleased with each other, floated through the bedroom door.

When all was quiet again, Grandma Marks came in with a large paper package which she unfolded. It was the floor plan for a generous sized house, three bedrooms and a bath upstairs; parlour, dining room and kitchen below.

"We be building this house next year in Langbank," she explained to her startled daughter-in-law. "After it's built we be quitting the farm and turning it over to you and Sam. Will now 'as a farm of 'is own nearby."

It took some time before Kate could realize the changes ahead for her family.

"We will be selling out and moving here, then? But Mother," she objected, "you should build a smaller house. That plan is much too large for you two. It will be too much work for you to look after."

" 'Tis for the future, Kate. 'T will be for you and the children one day."

CHAPTER 13

Kate spent blissful days resting. The children were quiet and well behaved with the individual attention they all received.

Mary liked to stand on a chair and overlook Grandma's baking. She questioned every recipe. Will frequently stood in a trance-like daydream, his hands tucked in the bib of his overalls, gazing across the prairie. Willie, his namesake, stood beside him and practised doing the same. Everyone petted Hazel and cuddled the baby.

Ruth toddled after her grandfather. He fetched the cows with their tinkling bells from the upland pasture at milking time, locked up the chickens for the night, and looked after the cats.

He sat down on a board nailed to a block of wood, and adjusted a tin milkpail between his knees. Leaning his head into the cow's flank he prepared to strip the cow of milk. Ruth sat on a smaller board nailed to a smaller block beside him, and watched the cats. Fat grey cats, yellow thin ones, big mother cats with kittens and several old Toms scampered into the cow barn as the first echoing thump of milk hit the bottom of the empty pail.

"How many cats have you got, Grandpa?" she yelled.

The cats ranged about the old man and the cow with little regard for the clicking heels of the milker, jostled with each other, cleaned their whiskers and meowed.

"Wye? Whuddee say, girlie? Whud 'ee say?" For every two streams of milk that went into the pail, a third found an open cat's mouth. From long practice, the old man's aims rarely missed their mark until all had at least one drink.

When the milk ceased to come down, he stood up, put away his milk stool and called his pets to a ring of pans nearby. The cats wound themselves lovingly around his legs, endangering the full pail of milk at every step, purring and yowling.

"Here, Sib! Come, little Sib! Come, grey Sib! Come, Mother Sib!" He clanked on the pans and then filled them with the warm milk.

"Why do you call them all Sib, Grandpa? How do they know which one you're calling?"

"Eh? Whud'ee say? Whud'ee say?"

The little girl, conditioned to her grandfather's hearing deficiencies, followed him from pump, to pen, to cowbyre, and shared a tacit understanding with him that most questions didn't need an answer anyway.

Later the children gathered about the kitchen table where preparation for making Devonshire cream was in progress.

"Bring up last night's milk from the cellar floor, will 'ee Ellen? And mind, don't jiggle it," the grandmother instructed.

"Tell us about Aunt Sarah and the Queen, please, Grandma," the children begged.

The old lady took the large, round pan of milk from the hired girl. A thick layer of yellow cream had surfaced on it. She very carefully set the pan on the back of the kitchen stove.

"Did Sarah like the Queen?" they queried. "Did she want to go to London to work in the palace?"

"T'was not if she liked her or no." She gathered the children close around her and sat down on a chair. "The Queen Victoria, God bless 'er, came to our village and 'eard tell of Sarah and 'ow she made Devonshire cream. So Sarah, 'er be told by Royal Command to come to the palace and cook for 'er Queen. A very great honour!"

"And did the Queen eat Devonshire cream every day? Did Sarah stay at the palace until she died?" The story, told many times before, was retold with relish while the clotted cream recipe was shown and explained. The fire geared down to a semi-glow.

An hour passed. The surface of the milkpan began to flutter and ripple.

The children shouted, "Grandma! Grandma! It's starting to winkle. It's winkling, Grandma!" The big pan was carefully removed from the stove, cooled and just as carefully carried down cellar again to be left on the earthen floor until the following morning. Then it was again brought upstairs. An egg turner was slipped between the top layer of cream and the thin milk below. The clots of thick cream were piled in a bowl and set aside for afternoon tea.

Several days of every week during the summer weather visitors came to spend the tea hour and visit.

"Your Devonshire cream is the talk of the district, Mrs. Marks!" Two ladies, properly turned out in veiled hats, gloves and long duster coats, had driven several miles with buggy and team across the prairies for a visit.

They sat in the comfortable living-room around a tea-table covered with snowy, lace-edged linen. Ellen replenished the tall silver teapot from time to time. Grandma Marks filled each eggshell thin china cup. Fresh mounds of hot buttered biscuits, spread thickly with wild black currant jam and topped generously with clotted cream were served to the guests.

A happy hour passed. The children munched crinkle-edged cookies and played quietly in the kitchen with Ellen murmuring at them occasionally. After much talking, laughter and exchanging of recipes, the buggy was brought to the door. The two ladies said their farewells and prepared to leave.

"A visit with you is a tonic. So happy to meet your daughter-in-law." The guests drove off in a swirl of dust and disappeared around a turn in the tree-lined road.

CHAPTER 14

The log house drew other visitors for another reason. It was Edison's wonderful new invention that recorded music.

One evening two young bachelors, well scrubbed and shaven, rode into the yard, stabled their mounts and presented themselves at the door.

"Father must approve of them. He's bringing out the gramophone," Grandma Marks whispered to Kate as she moved back and forth to press cakes and tea on the guests.

"It's a wonderful invention, Mr. Marks," enthused one young man. The old man gave a brief nod of agreement and proceeded with the ceremony of the gramophone. It was taken from the bedroom and set upon the dining room table, along with a carton of round cardboard cylinders which contained the circular records.

The gramophone was a square oak box with a steel top, on which a shiny cylinder was fixed. The grandfather wound up the spring by a handle on the side. A blue cylinder from one of the boxes was carefully dusted with a piece of chamois, and fitted on the steel cylinder on the top. He adjusted a huge bell, shaped like a giant morning-glory, to a circular steel arm that hung above the box, attached the horn by a rubber joint to the machine, and turned the switch. Everyone waited in silent wonder. Suddenly from the bell came the first thin, wavering bars of "Drink to Me Only With Thine Eyes".

"Play 'Uncle Josh and the Birthday Party' " begged Willie. The other children crowded round, each shouting their preference. Their grandfather methodically wound and resound the handle each time the music slowed and wavered. For two hours Irish melodies, Harry Lauder's Scottish songs, and the voice of Jenny Lind, the toast of all Europe, drifted from the big bell.

Grandpa Marks left all company conversation to his wife. She presided over the guests like a happy queen, with droll wit and sparkling conversation each time the records were changed. When it was time to go, the guests were bid a brief good-night, and the magic machine was put away carefully for another night.

The children prepared for bed. "When we're cleaning pipes tomorrow, Grandpa," willie announced, "I've got a good idea. Old Joe, our neighbour, learned me it. He cleans his pipes and it's easy, Grandpa. If you lend me some of the cats I betcha I could clean them myself."

"Do 'ee think so? How would 'ee do it, then?"

"I'd do it the way Joe does it. He shoves the cats in one end of the pipe and they come out the other. White cats are best. You've got lots of white ones, Grandpa. When they come out white at the other end, your pipes are clean."

"Doo 'ee listen now. 'er be frightful crool to cats, I be thinkin'." Grandfather looked serious. "I be thinkin' of a better way an' 'appier cats."

CHAPTER 15

The cats were happier the next day, but everyone else suffered by the time the pipes were clean and the new stove in place.

The following morning newspapers covered the floors, dusters and mops were at the ready, and Grandpa, Will, Ellen and the children were about to launch the affair of the pipes.

Grandma's eyes twinkled, and she whispered to Kate, "We'll just go for a walk, now, Kate, to give the hens som'at for their breakfast."

Willie balanced back and forth on his heels with his hands in his braces like his uncle Will, and examined things. His attention was divided between the stormy scene of battle in the picture on the wall, and the methods of procedure used to handle pipes.

"Joe always marks numbers on his pipes when he takes them down," he shouted up at his grandfather.

" 'Er be the one dredful crool to cats?" Grandpa stood on a chair above him and wrestled with two pipes stuck together and refusing to come apart easily. Will stood on another chair at the opposite end of the room supporting the pipes there.

"Yes, he's the one, and he always knows which pipe goes back first."

But the old man never heard. Suddenly the pipes jerked apart, he lost his balance and the chair skittered from under him. He flailed wildly to save himself, caught the support wires which immediately ripped from the ceiling. He came crashing to the floor, pulling everything down on top of him. The pipes all burst apart, falling one after another like the clatter of the battle-ground cannon-balls.

Each pipe trailed its own black soot as it rolled.

"Gor' darney all, Will. Can't 'ee see wot 'ees about? Don't 'ee know 'ow to keep pipes up? Grab 'em fore they mess the floor!" The old man maintained his sitting position in the centre of the room while everyone else scurried to mop up.

It took some time to remove the soot and take the pipes to the garden. The insides were cleaned with a duster on the end of a piece of stiff wire, and they were suspended again from the ceiling. Grandma and Kate were back in time for the next crisis.

"Time to move the heater now, Father." Grandma took the children to the kitchen to watch while she made Cornish pasties.

Pastry, cooled in the cellar from the night before, was rolled and cut out in saucer-sized circles. Raw beef cubes, potatoes, onions and turnip were placed on the rounds, folded over, and the edges sealed with a fork.

"Put my name on mine." Willie watched the procedure with the concentrated attention of a hungry six-year-old.

"Mine too, Grandma, please! Mine too!" and the other children pressed close while the pasties were popped into a slow oven.

Ellen, Will and the grandfather began to move the pot-bellied heater from the summer kitchen where it had been stored during the hot weather, and put it back up in the living room for winter warmth. It had a square base with curving iron feet that fitted on each corner, and it weighed like a ton of lead.

"Grab 'old, Ellen. 'Ang on, Will. 'Old tight to un now!" Grandpa shouted orders and backed and yanked, always in the way as the two helpers inched the unwieldy article towards the doorway through which it must be taken. They had it almost through when the whole weight swung around and caught Grandpa's hand between door and doorjamb.

"Owwwhh! Owwwhh!" The little old man let go his side to yank the hand free and before Will, red faced and perspiring, could save it, one leg fell off and the stove slowly descended on Grandpa's big toe.

"Will! Will! Gor' dairney all! Bain't 'ee any sense? Can't 'ee see wot 'ees about? Oww! Oww! Grab 'old Ellen! Grab 'old, will 'ee, Kate? 'Old! Owww! Owww!" He was freed at last and hopped about holding first his toe and then his hand, and rattling off Cornish curses on all cast-iron inventions.

"Why, Father," his wife admonished calmly, "do remember the children."

"He looks just like Rumpelstiltzskin, doesn't he, Mama, the way he hops about?"

Grandma and Kate rushed in with bowls of warm water, bandages and sympathy, not daring to look at each other as they passed.

The children stood about watching the happenings with wide-eyed wonder.

"We'll all 'ave some Cornish pasties," announced Grandma, "that'll 'elp ease the pain."

By the time the room was in order again and the winter heater in its proper place, the aroma from the kitchen oven permeated the rooms.

Later Grandpa, one foot and one hand in gigantic white bandages, sat on a sofa with the tea table drawn up beside him. His hot cup of tea and a large, warm pastie were comforts for his physical injuries, and the undivided attention of the whole family soothed his other hurts.

"Will your toenail turn black and come off, Grandpa? Will it hurt very long, Grandpa? Will you be able to feed your cats tonight, Grandpa?" The questions continued.

"He's reg'lar as Christmas," his wife whispered to Kate. "He lost the toenail on the other foot the very same way, at the very same time last year."

The way home, several days later, was long and wearisome.

"Now, children, don't quarrel." Kate handed out sandwiches and cakes from her well-stocked basket, as they jogged along. "Think of things you liked best at Grandma's."

"The pasties! I know how to make them, now," decided Mary.

"MMM, all those nice cats. I liked the cats best."

Uncle Will learned me to ride. I can ride a horse now." This from Will's namesake.

"Grandpa and the pipes," shouted Hazel. "Grandpa and the pipes." Everyone laughed, cuddled down, and fell asleep.

CHAPTER 16

"Jessie, could you come and give me a hand for a couple of weeks," wrote Kate to her sister the following summer. "We are expecting an increase sometime in August. The money would help you buy some of your uniforms and things."

Jessie, with a few weeks left before she entered nurses' training, arrived one hot summer day.

"Kate, do you think you should be sewing so much?" She watched anxiously as her sister piled another pair of boy's pants, cut from an old pair of men's overalls, on top of a growing heap of freshly sewn garments.

"And how can you stand to handle those woolen materials in this heat, anyway?"

Kate smiled happily and cut into a big bolt of blue serge. "This new Singer sewing machine is just a pleasure to run! And don't worry, Jessie I feel better this time than I ever felt before. Besides, look at the children's things I'll have ready for the winter. Ten dresses for the girls and as many things for the boys."

It was hot with that heavy, strangling heat that long rainless spells on the prairie can produce. The children wandered in and out or lay on the cool floor, heavy-eyed and petulant. Outside the sun, a copper disc, shone down with merciless glare. The leaves of the aspens behind the house hung silent and limp, and all the garden wilted.

The sewing machine whirred on, almost the only sound to break the thick stillness.

"Don't you think you should rest now, Kate?" Jessie's hair hung in wispy strands about her perspiring, anxious face. "It seems to me you stay at it too long — and all that bending over — it's not good ——"

"Don't worry, Jessie. I feel just fine. It's due the end of this next two weeks — I think. Mrs. McMillan said she'd come. She's better than a doctor, everyone says."

Kate took her foot off the pedal and leaned forward on the machine. In the silence, a faint but ominous rumble floated through the open windows. The room darkened and a slight breeze fluttered the curtains. The children, jolted alert, sat up and looked alarmed.

"That's plunder. I just hates plunder." Hazel cringed close to the sewing machine. Kate stood up, clutched her side for a moment, gasped and sat down abruptly.

The curtains, which had hung like limp rags, suddenly snapped out and streamed into the room from the rapidly darkening windows. The gentle breeze outside accelerated to a gale. Trees thrashed, swayed and beat against the side of the house. Rumbles of thunder and crackles of explosive lightning rolled over the housetop like a clatter of guns. Flashes of light reflected on the pale faces of the children. Faint pings from the telephone followed each clap.

"Close the windows, children!" shouted Jessie. The two older girls rushed to help.

Across the flat fields to the west a spiraling cone, deep black against the purple sky, danced and writhed and snaked along the horizon and then turned and approached with alarming speed.

"Will the cows know enough not to run under the trees like they did last year, and get killed? Will they?" Ruth looked into her mother's face, now greyed with pain. Kate seemed not to hear, and didn't answer.

"Lie flat on the floor, everybody, at once! On your stomachs and cover your faces! NOW!" screamed Jessie above the storm's roar. "You too, Kate —" She ran from one child to the next snatching tablecloths, clothing, anything she could grab, and covered the children. Kate was gritting her teeth and gripping the sides of the machine with sweat-slippery hands.

"Are you all right, Kate? Is it ——? Oh, God!" Jessie grabbed Kate's shoulders and tried to pry her hands loose just as a bomb-burst of sound thundered past outside. In a sudden vacuum of noise, the window glass exploded, buckling outward and carried away with the roar of the wind. Some fell inside and splintered over the floor and the children.

Outside, the cyclone ripped a wide path through the barnyard between the house and the stables, pulverizing fenceposts, pens and hen-coops, and showering down chunks of debris on the rooftop as it passed. For a moment or two there was a stunning silence, then the rain struck. It hit the housetop with the sound of guns and covered the blank windows with a solid grey sheet of water that streamed over the floors and flowed unchecked into all the rooms. The roar of the wind lessened and the rain, with a bombardment of hail, increased in volume and fury. Where the twister had passed, water formed, brimming up to the steps and pouring over the doorsill.

"Get into the girls' bedroom. Be quick, children! The North windows haven't broken yet. Hurry!"

The children scampered to the safety of the bedroom and crawled shivering beneath the quilts.

"Try to get to the bedroom, Kate. I'll call the doctor and get Mrs. McMillan." She helped her sister to her feet, rushed to the telephone, took down the receiver and cranked the handle.

"All my sewing — drat the luck." Kate ground her teeth and stopped halfway to look back on all her work, now sodden lumps floating in pools of water.

The rain drove in buckets through the windows, and poured in the doorway. Sounds of splashing pounded up through the floor boards below the house.

"Jessie! Jessie!" Kate shouted, "the water's pouring into the cellar!"

Jessie hung up the receiver and turned from the phone. Her face was drawn and her hands shook.

Kate swayed against the wall.

"My crocks of pickles — and the salted meat — the churning of cream — they'll all be spoiled. Save the cream, Jessie, for heaven's sake — the cream —" She grabbed the back of the chair and held on for a spasm's length.

"The cream! The cream!" Jessie's voice rose hysterically. "Never mind the cream! I can't get Mrs. McMillan. THE PHONE'S DEAD!"

Tear stained faces appeared at the bedroom door.

"Mama! Mama! What's the matter?"

"Get into bed and stay there. The storm's almost over. We'll be all right."

Kate set her chin grimly and turned to Jessie. "Babies have been born without doctors before. We mustn't frighten the children. Help me to bed, and quit worrying."

Time dragged on, the rain lessened and the occasional sounds of whimpering children were heard over the storm. Muffled exclamations and snatches of conversation came from Kate's bedroom each time Jessie rushed out to the windows to see if anyone was coming. A rumbling of wagon wheels and jingling harness announced Sam's return from town where he had outstayed the storm.

"Tell your father to get Mrs. McMillan at once," Jessie directed Mary. "And tell him to call the doctor. Hurry!"

The sun came out and sparkled coldly through the paneless windows and reflected on the small lake of water which covered the yard. Rain still dripped from the eaves and the steady plink of water falling came from below the house.

The tornado had cleaned the yard like a giant suction machine. Dead chickens and other fowl lay where the water had receded. The trees at the back of the house were splintered like toothpicks and the garden was washed away.

Mrs. McMillan bustled in and rushed to the bedroom, where she closeted herself with Kate.

Jessie ran back and forth with pans of boiling water and big pots of hot tea. The anxious children were sent to the barn to see if there were any eggs left after the storm. When they returned, wet and shivering, a baby's loud cry floated out of the bedroom. They huddled around the crackling cookstove to dry and warm themselves. Jessie opened the bedroom door, her arms about a roll of blanket from the end of which two very small blue feet protruded.

"Who wants to see their new sister?" she asked, pulling back the covers and exposing a red, wrinkled face for inspection.

The doctor, driving a pair of spirited bays, spun into the yard, threw the lines to Sam and rushed for the house. Later he sat beside Kate, sharing a cup of hot tea.

"Well, Mrs. Marks, your children are like their father. Speed, speed, speed! They can't wait to get here. Isn't this the sixth you've had without a doctor?"

"We'll call this one Edith, eh, Missus?"

"Only one name, Mr. Marks?" asked Mrs. McMillan.

Sam grinned. "At the rate we're going we'll run out of names before long. One's enough, anyway." he handed his new daughter back to his wife and went out to see what had survived the storm at the barn.

CHAPTER 17

At last the house in Langbank was built, the homestead in Poplar Grove turned over to Sam and the old folks retired to Langbank. The farm

on the Qu'Appelle was sold and the packing was finished. Grandpa Marks came to help the family move.

"Do 'ee ride with me, Kate." The old man helped her, with the baby in her arms, into his buggy. Buster was squeezed down at their feet. They set off southward at a good clip and were soon out of sight in the greyed dawn. Hard pellets of snow spattered down, drifted in the frozen ruts and quickly obliterated their trail.

Behind them creaked hayracks piled high with furniture. In the last wagon, loaded with bedding, the other four children huddled, yawning and cross from early rising.

"There won't be any buffalo bones to make barns and things with." Willie kicked the side of the wagon and looked back at their white playhouses being swallowed up forever in a swirl of whiteness. These structures had taken the children many days collecting bones scattered over the hills; bones from herds of long-dead buffalo, slaughtered in Indian forays long ago.

Sobs and sniffles began as the pretty green house on the banks of the valley disappeared in a cloud of whirling snow.

"Grandpa has made us a playhouse of firewood, and wood violets grow right close behind the house." Mary dried her eyes and tried to be cheerful.

"The flowers are all frozen," Willie accused, and turned his head to look back down the whitening trail.

"Well, Christmas will soon be here and there's a good sliding hill right near the house, and there's a nice spring with water bubbling up—" Nobody smiled.

"Let's sing, everybody! 'Jesus Bids Us Shine'," and the four young voices rose fitfully over the grinding wheels of the wagon as they jolted southward.

Late at night Kate unwound the two-month-old infant from her cocoon of covers, fed her and settled the two children in make-shift beds in the clean, empty rooms. Grandpa Marks soon had fires blazing and crackling in the stoves and the flickering flames, as he piled on the fuel, lit up the freshly whitewashed walls and the low varnished ceilings of the log house that would shelter the family for the next two years.

Babies cared for, Kate was ready to work again. "Oh, Father, I'm so glad to be here," she sighed. Loving thoughtfulness emanated from every spotless corner. Grandma Marks had purposely left the big coloured calendars and framed pictures for the children.

"Bide a bit, Kate." Grandpa rigged up a seat from a box for her by the fire. "I be warmin' these pasties and makin' a cup of tea for 'ee," and the old man bustled about preparing food while Kate sat and gratefully accepted the service.

Wagonloads of furniture arrived later and were hastily set in place by the men.

"There's a room for the girls and a room for the boys!" Mary and Ruth took off coats and undid shoelaces in the lamplight's rosy glow. Grandpa gave each tired child a giant Cornish pastie on their way to bed.

The wavering light from the fire twinkled on the shiny varnished ceiling of the bedroom. Willie lay in bed and looked up at the picture of the Scottish soldiers, "The Thin Red Line", on the wall. Grandma had remembered he liked it!

"Anyway, there AREN'T any buffalo bones in this place," he muttered stubbornly, but he smiled at Buster, already asleep, snuggled warmly beside him.

A year flew by. The farm prospered. Kate was happy and the children found the hills and trees, so close, a wonderful place for play.

After a mild winter, the first day of spring arrived. Saskatchewan, never influenced by the calendar, was wrapped in a roaring blizzard. Mrs. Lloyd, a British-trained midwife, had been in charge for a week, and the household showed it. The children tip-toed where they had clattered, and whispered where they had shouted. Every toque, mitt and coat had been trained in so short a time to hang on its own nail. Every bed was made adequately, if not neatly, before the cutter left for school.

"No school today." Sam tramped in, banged the snow off his boots on the side of the woodbox, and trailed across the room to check on his wife. "Too cold. I need the horse to round up the cattle in this storm. Don't want to lose any." He thumped back out, banging the outside door and leaving a trail of melted snow behind him.

"Mmph," sniffed the midwife. She scattered the children into their bedrooms and commanded them to be quiet. She swept up the snow, made tea and hot toast and took it to Kate in the bedroom.

"Just you relax now, Mrs. Marks." She set down the tea things and plumped up Kate's pillows. "By my calculations, the baby should be here in a week. A good rest in bed will do you good."

"I'm afraid you're a week late, Mrs. Lloyd." Kate sat up abruptly. "I think you'd better call the doctor."

"What!" The nurse looked startled and very annoyed. "I've never been wrong before!" She flounced out of the room and made for the telephone.

She cranked the phone repeatedly, took down the receiver to listen, and hung it up again. The wind increased. It moaned and tore at the house corners and ripped the shingles from the roof. Snow swirled against the windows and doors, wrapping the house in a white shroud. The children, alarmed at the storm and the sounds of anguish coming from their mother's room, opened the bedroom doors and peeked out.

"Back to your room, children, at once!" Mrs. Lloyd scurried in to the bedroom again. "The phone won't answer," she announced.

Kate walked the floor and wrung her hands, smothering the pangs of childbirth by gritting her teeth.

"I hope Sam gets here soon, with those cattle. It's not safe to be out in this weather."

"Mmph! Men!" Mrs. Lloyd snorted. She made her living from what she criticized most. "If husbands had the children, the race would die out in one generation!"

The day wore on. The constant roar of the wind was a terrifying sound. The nurse, more distraught by the moment, rushed needlessly back and forth from the sickroom to the telephone. Mary, taking advantage of her inattention, brought the children out to the warm kitchen.

"Come, children, and have some breadfood." She poured heated milk over chunks of dried bread and laced it well with brown sugar. Caught up in the anxiety surrounding them, the children sat with untouched plates.

"It's called Devonshire sooey," Mary explained. "Grandma taught me how to make it." She coaxed a mouthful into the baby.

The kitchen door burst open and Sam, bringing in clouds of snow and sleet, stomped into the room spraying the clean floor with more snow. His eyebrows and moustache were covered with ice.

"How are you, Missus? Got all the stock in," he shouted so Kate could hear.

Mrs. Lloyd popped out of the bedroom, closed the door behind her, and flew at Sam like at angry hen.

"Mr. Marks! I insist that you do something about this telephone — AT ONCE!" Pince nez askew on her sharp nose, she fixed Sam with her steely grey eyes and stopped him in the middle of removing his sheepskin coat.

"I've been calling Dr. Hilts all day. There is NO sound whatever on that line." She glared as if Sam was responsible for the storm, the telephone breakdown, and his wife's condition, as well as the snowy mess on the kitchen floor.

"What's this? What's this?" Sam, worried and angry, was about to answer in kind. At that moment a new sound rose above the roar of the wind. Still in his winter boots, he rushed for the bedroom, followed by the nurse.

Later he opened the door and a loud wail floated out.

"You've got a new sister. Now what will we call her?" Kate's seventh child, Amy, made her entrance into the world, as was usual with Marks children, without benefit of a doctor. The roads were impassable for days and he never did get in.

One day late that summer, word came that Grandma Marks was very ill.

"Mary, you take good care of the children and feed the baby this diluted milk. I'll take Edith with us. Call Grandpa's if you need us." Sam and Kate drove off to town.

Later, with her husband, sons and Kate by her bedside, Mrs. Marks Sr. died, still a young woman at fifty-eight. Her husband closed the house in town and went back to England on a trip.

Harvest work began and left very little time to mourn the gentle and beloved Grandmother. Sam had a bigger gang than ever to keep the new harvest machinery in action.

"I've a chance to buy some good land east of Langbank. What do you say to that?" He was doing his yearly bookkeeping. "There's a small house nearby, belongs to a neighbour, we can have for a while till we build. We won't have to move from here until spring."

"How far is it from town? The children need to be fairly close to school." Kate looked apprehensive. "I hope there will be good water handy." Schooling was important. Both parents had gone to work at thirteen, and were self-educated through reading. Sam read the daily newspapers, the Grain Growers' Guide and the Free Press Prairie Farmer. Kate read everything she could find.

Sam had a built-in arithmetic sense and could calculate wheat prices, wages, bushels of grain, and costs, with lightning rapidity.

"I hope that house will be big enough for this brood," sighed Kate. It was Saturday, and bath night for her obstreperous brood. "Do you know, Sam, we've had seven children in less than ten years!"

"What were you doing those other three years?" her husband answered, still busy at his desk with production costs.

CHAPTER 18

Kate fiercely denounced anything that smacked of snobbery, yet she was proud to the point of smugness of her cousin, the Rev. R. L. McDougall.

"Cousin Robert has the best education in the country," she told her children repeatedly. "He speaks seven languages — translates the Bible from the original Greek for his sermons. He was one of the most brilliant graduates of Glasgow University, and could have had a church anywhere in the big cities, but he chose to come here to south-eastern Saskatchewan in the early days and be a humble, itinerant preacher." She would shake her head, cluck in a kind of pseudo disapproval, with a little smile on her face.

"I'll get his diaries some day. We're his next-of-kin. We'll donate them to the Regina Archives, after we've read them, of course."

Reverend McDougall batched in a tiny granary-cum-shanty in the

Brookside District some miles away. During the winter he boarded with the Widow McLean nearby.

One early spring morning the widow eyed her charge with disapproval and some regret.

"Tis no time to be removin' your winter clothes, your Revrunt, and so it ain't!" You'll catch your death in these raw April winds!"

The steady prairie diet of porridge, potatoes, gravy, pickles and pie had increased his already ample girth. The widow's watchful eye now noted a certain sag in his suit which meant the time had come — time to take off his winter woolies and time to take off on his year's missions.

During the coldest months at the Widow's he kept a locked cupboard containing diaries, extra Bibles and his winter coat. The rest of the year he walked the prairie trails, ministering to the scattered Scottish settlements in eastern Saskatchewan. He preached in farm parlours, at dining room tables or around kitchen stoves, in one-roomed school houses or lonely little prairie churches. His sermons were always in the Scottish Gaelic first, for the older parishioners, followed by a repeat service in English for their sons and daughters.

Through the winter blizzards, he buried himself in the bare, cupboard-sized room in the corner of the old log house. He protected himself from the cold that seeped through the chinks between the logs by bulky woolen underwear, sweaters and vests worn under his black clerical suit. A long woolen scarf encircled his neck and covered his shoulders.

Above the roar of the wind, the steady drone of his voice reading the scriptures from Greek or Latin Bibles, or humming snatches of song as he translated his favourite hymns from English into Gaelic, or the scratching of his pen as he wrote endlessly in his diaries, were all comforting sounds to the lonely widow.

The first warm Chinook crept across the prairies, magically shrivelling snowbanks and dissolving the frost on the window panes into pools of water on the window-sills.

Mr. McDougall emerged from his room like a round black moth struggling into the warmth and light of spring. No return of winter weather (temperatures would sometimes drop sixty degrees in a few hours) could make him put on his underwear again, or, once he started out on his missions, make him turn back along the trail until he returned in late fall.

His silvery beard, untrimmed all winter and covering his cheeks like the soft folding of feathers, ended in a natural Vandyke at the point of his chin. Unmindful of the scolding, his round blue eyes twinkled, his round daub of a nose shone, and he beamed at the Widow like a happy child just awakened from sleep.

"You've been more than kind, more than kind. Spring is here! God be praised! God bless you, lass! God bless you, lass!" and the little man set about briskly sorting and packing for his long walks.

In a squashed, worn, leather satchel he packed a change of socks, some summer underwear, a red hymn book, his Bible, three bars of perfumed soap in a yellow box, and one large white towel.

A neat accumulation of orange-wrapper tissue papers (prairie Kleenex at that time) and squares of newsprint the same size, were rolled, banded with elastic and tucked in his hip pocket. As a concession to the widow's anxiety, he put on a scarf, a coat and a winter hat. From the cupboard he took a peeled poplar stick which he used to wave and thrash about him at the dogs that rushed out, barking furiously, from every farmstead.

Crows wheeled and cawed over last fall's bared stubble. The pastures were purpled with crocuses and every sidehill and ditch sang with running water. Reverend McDougall took the staff in one hand, the worn satchel in the other, and happy as the freed spring waters, set off on his annual visits.

He murmured to himself for the first few hundred yards down the trail and as the April sun warmed down on him, his voice grew louder.

"Mine eyes have seen the glory of the coming of the Lord——" he hummed. Gradually, with drumming feet and swinging arms, his deep voice, carried by the wind, rose and rolled over the prairie to the militant beat of his favourite hymn.

He crossed a bridge over a shallow ravine; in summer a place for wild hay, and now a brimming swirl of golden spring water that burbled along like a river. He set down the satchel, removed his clothes and hung them on bushes nearby and plunged into his favorite spring bathtub. With his back to the trail, he sat in the middle of the torrent like a happy pot-bellied Buddha, washing away with his special soap, pointed goatee sticking straight out and his eyes on the blue skies above him. He sang at the top of his lungs,

"Shall we gather at the river,
The beautiful, the beautiful, the River?
Gather with the saints at the River
That flows by the throne of God."

The water, bubbled and sudsy, eddied about him and then swirled off under the bridge.

Some miles away, two young ladies, properly hatted and gloved, decided to take advantage of the first sunny spring day to pay a call on the Widow. The old man didn't hear the sound of the buggy wheels or the clip, clip, clip of the horses' hooves till they thundered over the bridge. At the same moment he noted the passengers were both female. Without losing a note he continued to sing and wash while the two ladies held their breath, clucked nervously at their horses and waited until well over the next rise before they burst into laughter.

"Better not to notice! Better not to notice! Too late now! God be praised! God be praised! No harm done!" and he launched into "We Shall Have Showers of Blessings" until the sound of the wheels echoed only faintly over the hills.

Shortly after leaving university, McDougall fell deeply in love with a woman in the fashionable London church in which he began his preaching career. His forthright dictums as to what was sin alarmed the Bishop and McDougall was warned to tone down his sermons for fear of upsetting the generous and wealthy parishioners.

"I shall preach as God dictates, and not by the dictates of man!" he roared and so parted from his parish, his intended bride and his comfortable way of living. He took one of the first passenger trains to cross Canada west.

His small person emanated a kind of power which compelled men to quietly put away their bottles or clean up their language when his sturdy figure, coat-tails flapping in the wind and leaning heavily on his cane, appeared on the horizon.

Whether by design or natural consequence, he trudged the prairie trails, never quite knowing when the sun rose where he would be when it set. His most impassioned sermons were those concerning Jesus' wanderings in search of converts.

When he was presented with a silver collection, meagre as only a prairie hard-times church collection can be, he frequently gave it to the children of the house where he was staying at the time.

"The Lord will provide! The Lord will provide!" was his answer to any suggestion of censure, and gathering up his old satchel and cudgel, waving his hand for a blessing on the house, he made off down the trail. Behind him the congregation would mutter, shake their heads, and perhaps present him with a pair of home-made mitts, a second-hand overcoat or a knitted muffler the next time he appeared.

CHAPTER 19

"Dad's word is law around here," Kate often stated. She deferred to him as undisputed head of the house and saw to it that everyone else did too, until her cousin, Reverend Robert McDougall, came along.

"It was through Cousin Robert that I came west in the first place," she replied one day when she knew he was coming. The children begged to know why she had left the beautiful Muskoka Lake region in Ontario to live on these dry, lakeless plains.

"He said I could get work easily here!" She thwacked the great heap of elastic dough she was kneading in the bread-mixer with more vigor than the job seemed to require.

"You girls hurry up with your work now. He'll be here shortly." She paused in her bread making to pop a bottle into a howling baby's mouth.

"You boys get your wood and water in now." She shooed a dog or two out the door and put the kettle on for tea.

"Ruth, you get that flour-sack sheet I just made and put it on the cot up the backstairs for tonight. He'll likely be along today sometime."

"He's here now! He's coming in the gate. Gee, Mama, do we have to pray on our knees again? Do we, Mama?" Willie scuffed his boots against the empty woodbox.

"Just you mind your manners, young man, and go and get that woodbox filled, and close the door behind you. The rest of you clean up for dinner."

She smoothed her unruly hair, glanced for a second in the big, cracked mirror which hung over the sink, and grabbed a clean apron from a nail on the pantry door. Scattering young ones out of her pathway, she opened the kitchen door to the swelling sounds of canine altercation in the front yard.

"Children! Call off those dogs this minute!" Kate's tone had changed several degrees to "company calibre". She shaded her eyes with her hand and peered toward the centre of noise.

A big collie, one nondescript black mongrel and several tiny white fox terriers gave joyful tongue, jumping up and down, barking madly and wagging stumps and plumes of tails in a paroxym of welcome. In the centre, long black coat agape, clerical collar obscured by a flapping white scarf, a knitted brown hat with earlugs tasseling in the wind, stood Kate's cousin Robert. He waved his satchel and stick in the air and with loud shouts and warnings, advanced towards the house; all part of the game he played with dogs at every stopping place in his journeyings.

"Well, spring is surely here at last! We're so glad to see you!" Kate called off the dogs. A sharp gust of wind lifted and spiraled the dust about them. Robert tucked his stick under one arm, brushed the grit from his eyes, and held out his hand to his cousin.

"Ay, Ay, Katie-Cameron. God be praised! God be praised!" and the old man, puffing and muttering in Gaelic, followed Kate into the warm kitchen.

Once inside, he removed his coat and began to talk.

"Mrs. Willie Wickham has a son. God provides! A new pair of hands to help Willie on the farm. Wellie Stutt's son broke an arm. Fell off the barn roof. God be praised it wasn't his neck. Rollie Hoggarth sold his farm — going to the Coast. Ay, Ay! And Gordon Cole's team ran away — broke up the new buggy. Nobody hurt, God be praised! Hard time for the Moose Mountain people. Poor crop last year, but the Lord will provide! The Lord will provide!" The old man followed Kate about, talking and visiting until after lunch.

Then he sat in the rocking chair, put a newspaper over his face, locked his hands across his bulging waistline, and fell asleep.

"Outside with you now!" whispered Kate. She pushed the children out the door and quickly closed the downstairs bedroom where the baby was sleeping.

Robert slept — and snored. The children pressed inquisitively about the door until the first great snorting noise for which they waited, rumbled through the screen, a loud and glorious blast of sound.

"He's a better snorer than Grandpa," Willie squealed. Ruth shouted, "He's just like the Langbank train engine when it starts, isn't he?"

Kate suddenly and silently appeared at the door and the three children moved a short distance away so they could hear and not be heard.

The stentorian "Ho-oo tu-tut-tu tooOOOooo" rose higher and higher with each repetition. At last, with an explosive roar, it woke up the sleeper himself.

He got up at once and opened his satchel, joined by the children who pressed in closely so they could peer into the deep mysteries of the old black bag.

"Cousin Robert, you promised me the yellow soap box if it was empty, the last time you were here. Is it empty yet, Cousin Robert? Is it?" Ruth received her treasure, letting everyone have a sniff before it was tucked away with other treasures in her dresser drawer.

The old man took out his diaries, and unmindful of the noisy activity about him, proceeded to write and make notes for his Sunday service. Later he made preparations for supper.

He took off his suit coat, undid his collar, let down his braces and splashed copiously in the tin basin that sat outside the kitchen door. He dried on the white towel and with braces, collar and suit coat all brushed and in place, he tucked the towel into his collar in front and tied it behind his neck like a baby's bib. Then he took a big red cotton handkerchief from his back pocket where it always hung at the ready (except on Sundays, when it was replaced by a white one), and proceeded to blow his nose. This seemed not so much an exercise to clear the passages, as a sort of horn announcing "Grace". For, when done, everyone sat down at the table and bowed their heads.

Grace was a lengthy affair. He blessed the meat and drink, He blessed the crops they had and the ones to come. He blessed the whole family, including the hired man, and consigned them each in turn to God's eternal care.

As he continued enumerating each item, his voice began to run down and everyone stirred, expecting the cut-off. When the potatoes had cooled and the gravy congealed in the jug, he wandered off in Gaelic as if he was having a special visit with the Almighty who could only understand the true language of the Highlanders. Eventually, with a start, he ended abruptly and the meal began.

Supper over, in a silent signal, everyone, down to the smallest child, got down on their knees at the side of their chairs. Then the Reverend gentleman gave thanks for the meal which had just passed their lips, as well as the other bounties he'd forgotten to mention at Grace.

Unmilked cows mooed at a dry water trough, and pigs squealed hungrily down at the barn. Sam was getting restive as a fifteen minute sermon stretched to twenty-five. Every child wiggled and had had at least one crack on the knuckles and a fierce "Don't-you-dare-make-a-noise" look from Kate before the eventual wind-up.

The next morning after chores Sam, Kate, her cousin Robert and all the children in their Sunday best, set off in the democrat for the service at Earlswood, a Scottish settlement to the east.

Inside the tiny church the pews filled and the soft, quaint, half-quavering sounds of the Scottish Gaelic punctuated by an agreeing "Oh-I, Oh-I," murmured through the room until the Reverend McDougall entered the pulpit.

"We will take as our text today from Saint Luke, Chapter 9, verses 58 to 62: 'The foxes have holes, the birds of the air their nests, but the son of man has no place to lay his head'," intoned the minister, and the service began.

In the front pews sat a row of the elderly Hebridean settlers. Hands cupped behind ears, rocking gently or straining forward to catch every word, they punctuated the minister's every exhortation with the sound of keening and mumbled Gaelic blessings as the service progressed. A rousing sermon in the Scottish dialect was repeated again in English.

After church, the good man continued on his way, walking south from Earlswood. On the drive home the children were cross and hungry, and Sam was short-tempered. "He should cut it shorter — too long —" Kate said nothing.

"Dad, was Cousin Robert ever anybody's boy?" Willie stood on the back seat and leaned over Kate's shoulder.

"Yes. He was the son of an uncle of your mother's. Now sit down or you'll fall out over the wheel."

"Well, he wasn't telling the truth then. He said right in the church that the son af a man didn't have a place to lay his head and he does so. He can stay at the Widow McLean's, and so why does he say he hasn't a place to lay his head?"

Sam grinned and slapped the reins on the horses' rumps.

"Your mother will explain that, son," he said, happy that he wasn't alone in his criticism of the esteemed relative.

CHAPTER 20

That year, autumn began in late August, with the first touch of frost, and promised a bumper crop. Gentle winds rippled across the land, turning the wheat into seas of gold which swayed and dipped and waved as far as the world's rim. Bushes and knots of young trees produced wild red raspberries in summer and later high bush cranberries. Purple and blue saskatoons and chokecherries hung like grapes from bush and tree.

Havester "excursions" poured trainloads of men, European or "easterners" into the west. One man could seed a section of land, but it took ten more to help harvest it. The railroads provided cheap fares for the harvesters in the fall. The plan had benefits to the railroads as well as the farmers. These same harvesters often stayed on the prairies and acquired land of their own, or went home and brought back their families to settle in the west. The resulting produce, wheat, helped fill the coffers of the railways, and so the harvesters were an excellent investment in the future. Sam made frequent trips to town, sizing up the men who hopped off the trains, looking for good workers.

"I got two good men today!" He rushed in late for his morning snack, threw off his leather work gloves, pushed a straw hat from his perspiring forehead, and took the mug of tea Kate had quickly poured for him.

"I was lucky, I think, to grab these two. Both army men and used to handling horses. One's French and the other German." Sam began to laugh.

"You should see the German. He's still wearing his army uniform — high leather boots and all — like a dude general —" he laughed again, swallowed the rest of the tea, and grabbed his gloves.

"Dude or not, you should see the way he handles those horses—" At that moment, the subject of the conversation swung past the kitchen door guiding a team with the practised skill of a top cavalryman. The children ran out and stared with interest and admiration. The new man was tall and fair, with wide shoulders, and walked with the light step of a dancer. He wore a peaked military cap, sagging at both sides, and an immaculate grey-blue uniform, military badges still up. His high leather leggings shone with polish. He waved to the children as he passed, still guiding the horses with the multiplicity of reins.

"Lucky I don't get any of those I.W.W.'s from the east. I figured we'd be safe with these Europeans." And Sam was gone, letting the screen door bang back into place.

"What are I.W.W.'s?" Willie, always ready with questions, looked up at his mother.

"They belong to a society —" Kate hesitated, "Well, we call them "I Won't Work's", but their real name is Internatonal Workers of the World. We don't want them around because they won't work and they try to stop others. Dad's pretty careful choosing the men but you can't always tell. Last year he hired one by mistake and then fired him. They found stones and iron bars in the sheaves that could have broken the threshing machine and stopped the operation."

Noon brought the men in to dinner, a silent meal in which the children made careful inspection of the newcomers, Bruno the German and Reynaud the Frenchman. The contrast between the two men was startling. Blond, beautiful Bruno sat erectly and ate with finesse, answering politely when food was passed, and smiling at the children.

Reynaud, who had blown his nose on the doorstep earlier by the handy prairie method (closing one nostril and then the other with thumb and forefinger and blowing) bounced into the table on his short bandy-legs as if strung by wires. He yanked out a chair and sat all through the meal grunting when he wanted anything, and hunkered possessively over his plate as if afraid of losing it. Spikes of black hair shot out of the top of a skull elongated like an egg on its side, from whose narrowed ends two sagging ears suspended. A short brow was creased already with prairie dirt. His small eyes glittered on each side of a short, pointed nose and his chin, beneath a slash of a mouth, was almost non-existent. He sniffed.

When he finished eating, Reynaud banged the screen door behind him on his way to the barn. Ruth sang out, "Pop goes the weasel! Pop goes the weasel!" and danced to the doorway after him. "He sure looks like a weasel, doesn't he?" The children crowded by the door and watched him hurry away. "He kind of flicks along like one, too." Noses pressed to the wire screening, they watched the erratic gait of the little Frenchman.

"Shame on you!" Kate grabbed Ruth and swatted her bottom. "Get at those dishes, young lady, and don't you ever let me catch you making fun of anyone's appearance." She hurried the other children out of the door.

"Don't any of you dare say anything about him. You listen to him. Handsome is as handsome does. He speaks German as well as French and English, and he's an excellent worker. Never you mind what he looks like. That doesn't concern you."

Sam was on his feet swallowing the last drop of his tea and he caught his wife's eye as she turned to her work. He winked at her. "Pretty good description — good job he can work," he whispered to her as he swung out of the door.

The children were never easy with the Frenchman. They shrank from his touch and didn't trust his glittering eyes or the way he snarled at them when their parents weren't around.

Bruno, full of light and laughter and love, was the talk of the country and the family because of his apparel.

"My, your clo'es are nice. Did you buy them in Eaton's catalogue?" Ruth shared Bruno's genereous lap with Hazel and they both patted the strange material in his jacket with admiration. Mary, almost ten and too old to sit on men's knees, stood by with love in her eyes for the blond, young giant.

Bruno threw back his head and laughed, "And your teeth are nice, too." Hazel observed him from below. "You don't have tobacco juice on them like all the men."

Bruno laughed again and hugged the little girls.

"No tobacco," he said, "and no Eaton's. I get clothes from German government — present for work —" and he roared again at some secret joke.

"Better come to town and get some overalls." Sam looked doubtfully at Bruno's well-brushed and immaculate appearance. "You'll ruin that outfit. Keep it for Sunday. You might get in trouble wearing an army uniform — isn't it against the law?"

"I own! I wear!" Bruno stated. "I have two more for Sunday." Then he explained what had happened.

He hated the army in Germany, and with a friend had constant trouble with his commanding officer. After months of duty without a break, he had been granted a week's leave to visit his home. Instead, he packed his uniforms, crossed the English channel and slipped into a crowded harvester's special boat bound for Canada. He came straight to Saskatchewan without being interrogated or detained in any way.

"We fix officer before we leave," he laughed after describing his adventuresome escape. "He often beat my friend. Hit him with whip for noddings. We fix. We take his saddle, bury it in manure pile when we leave," and he roared again.

Reynaud, an ex-army man himself, hated Bruno and was jealous of his popularity.

"You should be reported wearing soldier's uniform. I report you myself!" Reynaud snarled at him one day.

Bruno laughed good-naturedly.

"You go tell Kaiser, hey?" He walked about the farm or into the village with the purposeful stride of a free and happy prince in his well-tailored clothes.

Indian summer came and everything sang with colour. At night the moon, big as a ripe pumpkin, rose and hung cloudy red-orange on the dark horizon. In the morning, blue skeins of mist trailed across the hills and frost touched the woods with yellow fire. The sun in the afternoon was a dull copper from the forest fires which burned in the mountains to the west.

Down in the valley a patch of chokecherries hung in great purple clusters, ready for picking.

"I just love chokecherries. Do you think the Indians liked chokecherries too, Mama?" Mary asked. She was helping get the children ready to go berry-picking.

"Yes, it was important for their health. They killed the buffalo and then hung the meat to dry and when it was dried they pounded in the chokecherry juice like Dad puts salt on the winter meat. The berries gave the meat a good taste and kept them healthy in the winter. Now give the baby a bottle in another hour and rock the cradle if she cries. Fry some potatoes for the men if we're not back in time — and some eggs.

Kate tied a red bandana handkerchief over her hair. She wore a loose "Mother Hubbard", a garment high-waisted and loose about the middle, the standard uniform of the pregnant women, and a hangover from her last pregnancy. She picked up two milk pails and arming each child with a lard pail, they set out towards the berry patch.

"Chokecherries make you mouth all puckery and you must be sure to spit out the stones too, or you'll get 'pendicitus," Ruth instructed, guiding Hazel over branches that tripped and stones that stubbed toes. They both toddled along in Kate's wake.

"I don't like chokecherries! This here's a girl's job. I just like horses!" Willie kicked at stones as he lagged behind the others. Buster, stumbling along on short legs, bellowed loudly and tried to keep up.

Kate paused on the trail and shifted the pails to one hand. She pushed back her hair from her white, perspiring face and felt ill and impatient.

"Take your little brother's hand!" she scolded. "You'll like the jam on your bread well enough, the dear knows, when it's made."

The laden trees were in sight and everyone pushed against branches and underbrush to reach for the hanging fruit. Tins filled up and mouths became black and puckery. Buster curled up at the base of a tree and fell asleep.

"Don't go too far, children. Stay within the sound of my voice. Kate anxiously checked and re-checked each child's whereabouts every few minutes. Soon the big pail was full.

"Willie, you take this pail back to the house. Carry it carefully, mind, or there'll be no jam. Bring back the empty one. Hurry!" She carried the berries out of the bush, handed them to her son and guided him onto the well-marked trail to the house.

She pushed back into the bushes, sat down by the sleeping boy and wiped her face. Ruth emptied another pail into the big one and went back for more. Kate arose wearily and set to work again. All was quiet in the trees except for the steady thump of the berries as they fell into the pail.

"Where's Hazel?" The continued quietness was unusual. "Hazel! Hazel!" Kate shouted. She grabbed up Buster, and forced her way through the brush to the trail.

"I'm here, Mama, and I've found a nice kitty to take home," piped a tiny voice from around the bend in the trail. Hazel waddled along the path towards them. In front of her stepped a black and white animal, its daintily undulating tail swishing behind.

"A tchunk! It's a tchunk! Keep away from it!" shouted Ruth, waving her hands in the roadway.

The warning was too late. The skunk was not alarmed at the child, but became angry at sight of the crowd. A haughty turn, with tail elevated, and she showered everyone with musk. The offended animal disappeared into the bush. The children all gasped and began to cry. Kate joined them.

When the men came for dinner the house was in chaos. The odour of frying potatoes mixed strongly with the odour of skunk.

The bathtub full of water was sitting in the back kitchen. Clothing was strung across the floor in every direction. Kate knelt on the floor, her face tear-streaked, her hair falling down on her shoulders, as she tried to calm Buster while scrubbing him vigorously.

The big German strode in and, hands on hips, surveyed the mess.

"Missus, you give children's clothes. I fix." Kate dropped the cloth, got up and went outside where she was promptly sick. Bruno took over.

He climbed down the cellar and hoisted a large stone crock, filled with buttermilk, onto the kitchen floor. With no ceremony, he emptied the washtub, half-filled it with the buttermilk, and put each child in turn into it, washing them from head to foot. Then he rinsed them off outside with rainwater from the barrel. The clothing was taken to the well and washed thoroughly in the steel trough. Then he piled each garment into the buttermilk and left them to soak. Later the clothing was taken to the garden and covered with a light layer of soil.

"You dig up in three days and wash, missus," he instructed.

Kate went through the cleansing process by herself and too ill to care, went to bed. The chokecherries Willie had brought to the house still sat of the kitchen shelf.

"I fix." Bruno cleaned the fruit and put it to cook on the back of the stove. When done he poured it into jars and put it to cool in the cellar.

The next day the men came in for a late Sunday breakfast. Sam looked harassed.

"The Missus is sick this morning," he apologized for the absence of breakfast.

"I cook then." Bruno towered over Sam and took over operations. He called the children.

"Ve make burble bancake. You fix table. I fix cooking."

Half an hour later mounds of big yellow cakes dripping butter and floating in a plate of purple syrup were set before each hungry man and child. Bruno, with a dishtowel floursack apron covering his immaculate uniform, sat squashed between Mary and Ruth.

"Bruno, I'm going to marry you as soon as I grow up," Ruth announced, between mouthfuls of food.

"Me, too, Bruno. Me, too!" shouted Hazel from her high chair. Mary put down her spoon and looked as if she were about to cry.

"Neffer mind. I marry all," and the big man encircled all the little girls in his arms.

CHAPTER 21

Indian summer hung on late that year. Days were still and yellow leaves hung grimly to trees stirred only occasionally with a gust of wind which threatened of cold to come. It was noon before morning mists cleared away. The wedged formations of wild geese honking in rhythm on their southward flight, flew over and only the whir of their wings and their wild song broke the long silences.

Reynaud boarded a train for the west. The crowds of men and the excitement of threshing time were past. Only Bruno stayed on.

"The Missus is sick," Sam apologized.

Bruno, washing at the sink after two hours of morning chores which included feeding the stock and cleaning barns, seemed not to have been touched by any farm grime. His uniform was clean and brushed, and his tall leather boots were free of barnyard filth. He dried on the roller towel and took the frying pan from Sam.

"I fix. You take care of Missus."

By mid-morning the kitchen shone with cleanliness and a big stew bubbled on the back of the cookstove. An ominous moaning drifted down the hall from Kate's room.

"Come, keeds. Ve haf picnic." He lifted Buster to his hip and cradled Edith, expertly wrapped in a blanket, in the crook of his arm. The big man stepped out of the house and swung down the prairie trail like a militant pied piper. A wild, happy, German marching song rippled from his throat in a glorious tenor which echoed and re-echoed across the valley and through the trees. The children, tripping over each other's heels, followed, silent and listening in wordless delight.

They stopped on a hillside where they had played all summer.

"Ve sit!" and everyone sat in a circle.

"First ve eat. Then ve learn. Eh?" Bruno produced chocolate saved from his trip to town, and divided it fairly for everyone.

"Now ve learn to count in German."

"Eine, swien —" Mary struggled with the gutteral sounds and they repeated it in chorus.

"Good, Good! Now ve sing—"

At that moment an odd staccato sound, unfamiliar and alarming, wafted over the fields. Everyone quietened to listen.

"It's not the steamer with the threshing. They're all finished." Willie strained to hear.

"I know," he shouted, jumping up, "it's Markle's automobile. I know it is. C'mon kids. I saw it in Kennedy last week." They raced off towards home, leaving Bruno and the smallest children to follow. As they rounded the bend in the road, a shiny black car chugged through the archway of branches over the lane and came to a snorting stop in front of the kitchen door. Two men in caps, goggles and mackintoshes stepped out of the car and disappeared into the house. The children shrank back in wonder and waited for Bruno. Then they surrounded the new car and climbed all over it.

"Look at the darling little lamps. Are they made of gold, Bruno?" The brass lamp standards, attached to both sides of the front, and brass trimmed fenders and hood blazed in the reflected sunlight. The buttoned-down black leather upholstery gave off an odour of newness, strange and exciting.

The children rubbed their hands over every surface, bounced up and down on the seats and took turns jumping off the rippled metal running boards that ran along each side of the car body. Willie discovered the horn, a big rubber balloon, and pressed it with startling effects. A great roaring snort wailed out, which caused the fowl in the yard to fly squawking into the air, and the children to jump back in alarm.

The kitchen door opened and the two men and Sam emerged, carrying Kate, who was wrapped in a blanket and didn't seem to see anyone.

"You children behave and do what Bruno tells you." Sam, white-faced and stern, climbed in the back seat beside his wife. One man stood in front and wound the crank which hung from the radiator of the engine. The car began to chug. It wheeled back onto the trail and rattled around the bend, disappearing in clouds of dust. The children stood quietly and watched it go, moving nearer to Bruno. Suddenly Hazel raised her voice in a wail of protest.

"Come, keeds. Effrybody help. Ve make bancakes and syrup again, eh?" and the ominous feeling that hung over the mysterious departure dissolved in the activity of helping Bruno cook.

In a western town some hundred of miles away, Jessie, now a nurse, raced down the street, pushing past startled shoppers and children at play. In one hand she gripped a satchel into which she had jammed a change of clothing, and in the other she clutched a worn purse. Her nurses' cap flopped loosely and hung to her mound of chesnut hair by one pin. A cape over voluminous starched skirts billowed out behind her as she ran.

A terse telegram had been relayed to her some minutes before.

"Kate very sick, come at once." She hurried to catch the one train going east.

A block from the station she heard the sickening sound of the passenger train gathering speed. She raced on, only to see the caboose rapidly growing smaller as the steaming locomotive roared off down the rails. Somehow she must stop it. She bounded up on the platform and made for the wicket.

"You've got to help me. I think my only sister is dying. You must stop that train. I must get to Moosomin!"

The startled station-master, peaked eye shade pushed up on his forehead, and with brown and black sleevelets covering his shirt sleeves, peered out of the window at a desperate, determined young woman. Her scarlet face was tear-stained, and glasses swung from a gold chain on her nurses' bib.

He shook his head slowly. The receding train was as final as yesterdays past. Jessie sagged against the wicket gasping for breath. Silently he directed her to a seat inside, tapped out a few notes of Morse code on his telegraph machine, and listened for a moment.

"No passengers again till tomorrow." He tapped some more and listened.

"Freight in ten minutes. They'll take you."

The next train through slowed for a brief moment and was off again. Minutes later a brakie, sneaking back for a smoke, was startled to find a young woman in crisp nurses' uniform sitting ramrod straight in the train caboose, a strictly male area, as it swayed and lurched on its way east.

The matron of Moosomin hospital was a short, rounded, black-eyed little woman, recently arrived from Germany.

"Iss hope, Dokktair?" She helped Dr. Wark to don his white smock, and moved swiftly about the sparsely stocked operating room preparing surgical equipment for an emergency.

The doctor shook his head. "Three days? Too long, I'm afraid. Never heard of one surviving, and least of all out here. But we must try. Tubal pregnancy. I've alerted the other doctors here and down the line. Might get some help."

Hours later Kate was bundled into a ward with one other woman. It was night. Bells whirred softly and constantly, and swift feet echoed along the uncarpeted corridor in answer. Across from Kate a husky young woman, convalescing from a recent operation, sat up in her rumpled bed and glared angrily at the anxious nurses as they came to check Kate's progress.

"I'd like a glass of FRESH water for a change, and bring me the bedpan AT ONCE!" she shrilled.

"Shshsh —" cautioned the nurse repeatedly, supplying what she wanted to keep her quiet.

Kate never heard. She lay in a coma. Her breathing scraped slowly from her labouring lungs and her bloodless face was as unlined as a child's. The thin thread which held her to living stretched towards the breaking point.

The night grew old. Doctor and nurses moved in and out on silent feet, checking Kate's pulse, massaging her cold hands and shaking heads in wordless anxiety to each other as they passed. Death, that grim spectre, who always hovered about the halls at night, in the grey confusion of dawn, slipped into the room, snatched and made off with a victim as the sunrise split the eastern sky.

"Miracle! Miracle!" Nurse Denkar held Kate's limp hand and felt her pulse, then whispered to Jessie, "Iss stronger! Iss stronger!" She rushed to tell the Doctor who sat nodding in a chair after a long, sleepless night. The angry young woman in the other bed lay still and lifeless. Kate lived.

"I heard them whispering in the night," Kate told her family later. "Someone said, 'What a shame to leave all those children motherless. Too bad it wasn't the other one.' I knew they were talking about me, and so I just gritted my teeth and made up my mind to live. It was the first successful operation of its kind to be performed there — five doctors — but it was Dr. Wark, your Aunt Jessie and Nurse Denkar who saved my life. Jessie special nursed me until I got out."

After the operation, winter settled in. Kate, pale but triumphant, arrived home in the sleigh box weeks later.

Spring brought many changes. Sam sold the Poplar Grove farm and moved the family three miles east of Langbank. Bruno helped move them into the tiny, cheerless shack by the new railroad, and then left for British Columbia.

Granpa Marks unexpectedly arrived home from England in late May, and telephoned as soon as he got off the train.

"Kate, I be comin' out and bringing a zurprise for 'ee."

CHAPTER 22

"Get a move on now, or you'll be late for school. Don't dawdle! Mary and Ruth, you keep your pinnys clean today. Don't go puddling in the water on the way home. Take care of Hazel now, and come straight home from school." Kate grabbed a book from her eldest daughter's hand and gave her a shove.

"Drat that child! Can't you see how late it is? Hurry up!" The clink of harness and the scaping of buggy wheels sounded outside.

"Here's Willie with the horses at last. Hurry up, now!" As the buggy wheeled out of the yard and disappeared over the hill she leaned against the door jamb wearily. The edge of the hill where the buggy had disappeared was a wavy, indistinct line before her eyes. Suddenly she grabbed at the door knob to steady herself as a wave of nausea rolled over her.

The squat, four-roomed house was inadequate for the growing family. There were no closets and no cupboards for storing things. Outside, the bare ugliness of the treeless fields depressed her. The area, subjected to prairie fires a few years back which had denuded the plains, had only a faint line of brush far to the north to indicate any tree growth. The two steel

tracks which ran along the south side of the yard glittered and cracked in the heat as the sun bore down.

"It's going to be a scorcher today — why did I agree to come before a decent house was built?" She slumped to a chair and looked back on the cosy comfort of the well-arranged rooms of the log house in Poplar Grove with regret.

Breakfast dishes sat unwashed on the littered kitchen table. Buster and Edith were shouting and playing horse with broom handles. Amy sat in the middle among scattered boots and broken toys, and howled lustily.

"Stop that racket!" Kate's shout was almost a shriek, and the children stopped in surprise. Even the baby quit crying for a moment.

"Oh, it's not your fault!" Angry and sick, she stood up and surveyed the jammed mess which she must attack at once.

She looked at the new green grass and the clean yard outside, and willed the sight to lift her spirits.

"You take Amy by the hand, you two, and go and get the eggs. The barrel of ashes Bruno had fixed outside at the corner of the house caught her eye. Perhaps being outside for a while might make her feel better.

"Grandpa coming, or no Grandpa coming, I'll make that soap today. A good day for it." She got a pail of water from the rain barrel and poured it over the ashes in the half-filled barrel, watching while it dribbled into the tub she had put below to catch it.

"Keep inside the yard now, because the bull is loose," she cautioned the toddlers, "and don't go near this tub. It's the lye for my soap." The children continued on their way. Amy wobbled along slowly, her legs as yet unaccustomed to bearing her whole weight.

Kate threw pail after pail of water over the ashes, and got her soap-making equipment ready. Out in the yard, a huge iron vat shaped like a giant's mixing bowl was set on top of a circular iron shield, under which a fire was set at pig-killing time. The freshly killed pig was always immersed immediately in boiling water and scraped, the pig's bristles much easier to remove at that time.

Kate piled kindling under the vat, then chunks of wood, and set it afire. The big bowl was an ideal place to boil her soap, and it would eliminate any further confusion in the kitchen.

"What was it I heard of that would make the soap firmer?" she mused. She made for the kitchen and took down the telephone receiver.

"Get me Mrs. Cairns, will you, Central?"

The three children reached the henhouse. Buster dropped the baby's hand and plunged into the centre of the hens and roosters quietly feeding or sitting on nests inside. The fowl, alarmed by the sudden attack, squawked in fright. Feathers filled the air, hens flew wildly into the wire windows or out the door, sailing over the frightened children's heads. The two little girls cowered against the wall until all the hens were safely outside.

"Hens won't hurt you, scardy cats!" Buster put the eggs he found into the tin jam pail and sauntered off to the stable, whistling through his teeth.

Inside the kitchen Kate stood with poised pencil, prepared to write down directions.

"I'm making soap today, Mrs. Cairns, and wondered if I had the right amounts. My soap always seems too soft to make into bars." She paused and listened.

"Mmm, I've got the water through the ashes to make the lye, now how much grease did you say to put in per pail? Yes, I see . . ." The conversation continued for some time.

"Sam's father is back from the old country on the train this morning and he's coming out. I'd better go, and thanks so much." She hung up, and realized that no sounds were coming from the yard.

She rushed to the door in time to see Buster prod the red bull through a sagging two-strand fence. The animal's glittering little eyes were directed towards the tiny nuisance on the other side of the wires, and he pawed the ground with a snorting bellow, lowering his head.

"Drop that stick and come up here at once!" Kate sprinted across the yard toward them.

A loud splash and screams following came from the side of the house where her soap was being made.

"Run, Buster! Run for the house!" she shouted, and not waiting to see, she wheeled and made for the sound of the screams. Amy was headfirst in the tub of lye water and Edith stood by the side screaming lustily, yanking at her legs.

Kate grabbed the sputtering child, dashed to the rain barrel and stuck her headfirst in the water. She pulled her out, struggling and choking, held her upside down and thumped her on the back. Then she immersed her again and forced her eyes open under water as best she could.

"Get the castor oil! Quick!" she shouted to Buster, who had escaped from the bull and raced up to join the excitement.

Kate repeated the process, with the baby gasping and fighting for breath between immersions. As she pried open the child's eyes and poured in castor oil, she was aware of the sound of buggy wheels close by.

"You know how to phone. Go and tell Central to phone for the doctor — AND HURRY!"

Buster raced in to do what he was told and Kate kept repeating the water immersion, the slaps on the back and the castor oil treatment doggedly until all the oil was gone.

Inside the house, the little boy pushed a chair over to the telephone, turned the little handle on the side of the wall phone, and unhooked the receiver.

"I'm Buster Marks," he yelled into the mouthpiece, "and my mother wants you get the doctor. The baby fell in a tub of lye water!" He hung up the receiver without waiting for an answer.

Flies buzzed over the congealed remains of breakfast.

"I'll just make myself some bread and syrup." He pushed the chair over to the table and reached for the pail on the shelf above. It slid away from him, spun out of his hands and crashed on the table of dishes below. The lid flipped off and a river of syrup gushed over the half-empty porridge plates and down onto the floor.

Outside, his grandfather's buggy pulled into the yard. The old man sat bolt up-right on the seat, his christy stiff hat at a rakish angle and a gold-headed malacca cane at his side. Beside him sat a tall, fashionable woman. Her wide-brimmed lace hat was swirled about with white netting and soared a foot above her head. Her well-corsetted figure, swathed in grey lace and topped by a coat made of silky green material, was helped tenderly down from the buggy by the old man. Then he tied the horses to a fence post and the two fashion plates made their way towards the house.

Kate's long, tangled, wet hair clung to her head on one side, held there by one remaining hairpin. Her soiled mother hubbard dress was soaked and clung to the swelling contours of her body. Two children, one at her feet and the other choking and struggling in her arms, screamed incessantly.

"Be 'er drownin' 'er childrun?" The woman, almost a foot taller than the grandfather, drew back in alarm and cringed against his side.

For a moment Kate forgot her half-blinded child at the sight of the visitors, and thought of the appalling mess inside the house. Then she screamed in relief.

"Father! Father! Call the doctor and see if he's coming. The baby fell into the lye water!"

Some time later the doctor peered into Amy's eyes.

"As far as I can tell, the eyes have suffered no damage, Mrs. Marks. You did just the right thing."

Kate's careless attention to recipes for once had proven lucky. She had added six times as much water to the wood ashes as was called for. The child's sight was saved by the diluted water.

Later Grandpa Marks, an apron around his middle, was finishing the lunch dishes and clearing up the kitchen. Kate and the children were all washed, dried, brushed and had a change of clothes. The tiny shack was as clean as the old man and the broom could make it. Kate finished up the dishes and got the doctor on his way.

On a chair in the other room, almost ignored in the excitement, the pale, trembling woman, Grandpa's new bride, huddled in one corner. Her big blue eyes dodged from one side of the room to the other like a cornered gopher looking for an escape route, and she clasped and unclasped her white-gloved hands.

The children pressed around her, putting out sticky hands to feel the lace of her dress, and staring up into her face. She drew back and pulled in her skirts.

"You're our new Grandma?" Buster stood in front of her and faced the possibility like a man.

The new bride suddenly leaned forward and whispered to him, "Be ther eny Indians 'ere abouts?"

Called upon for information, Buster felt it was his job to entertain and inform the neglected guest. He put his hands in his pockets and rocked back and forth on his heels like the men did when showing off their crops.

"Injuns? Oh, there's lots-o-hundreds of them — just on the other side of the track."

He stopped to think preparing to make it more interesting for his trembling audience.

"Yer safe in the day time. It's at night — and the big wolves. They howl something terrible. And then the Injuns all get around and light their fires — and then they dance around the fires. They always dance when they want to scalp people. Scalp is cutting off yer hair with your skin on it. Hurts awful. Don't you have any wolves or Injuns in England?"

With such a receptive audience, Buster expanded. Tales of badgers as big as cows that you couldn't kill unless you hit them on the end of the nose, and screaming wildcats as big as bears that were heard to come down chimneys at night if they were hungry, were embellished in blood.

The face of the bride took on a deeper pallor. She leapt to her feet when her husband and Kate came into the room prepared for a visit. She half-pushed, half-dragged the puzzled man towards the door. He was most surprised at his wife's undue haste, and made his apologies.

"We best be goin', Kate. Tell Sam to come in an' see me, will 'ee?" and they were gone.

"Well! So that's the surprise. Some surprise! I wonder what Sam and Will will think of her. An odd woman. A very odd woman. I wonder why she was in such a hurry?"

The younger children had gone for naps and only Buster remained awake. He was intent at Kate's elbow, and acting very much the man of the house.

"Guess she's askairt — I think —"

"For heaven's sake, what have you been telling her?" Kate paused in her pleased admiration of her now neat house and turned to her son.

"Well, I guess she don't like Injuns much — or wolves howling —"

CHAPTER 23

It was midsummer. Sam rushed in for breakfast, followed by the hired man.

"It's going to be a scorcher, Missus. They'll be two more men for dinner. They're helping me fix the barn this afternoon." He had his breakfast and was off out to the stable-yard, at a trot.

"He never seems to feel the heat at all." Kate watched him from the door for a second. "I wish I had half his energy." She turned to the muddled mess in the kitchen.

"Girls! You get at those dishes at once, and Willie, get that woodbox filled. Heat or no heat, the washing has to be done this morning."

The dirty clothes were piled in heaps on the floor. Milk pails stood unwashed beside the new milk separator. She straightened up painfully from time to time as she tried to bring organization out of chaos.

"Be careful with that separator, girls, and don't mix up the discs, or we'll lose half the cream again." The ugly black iron pedestal on which sat the round steel milk bowl, reeked of stale milk from recent spillings.

"It will save all that extra washing of pans and finding a place to put them," Sam had enthused when he dragged the infernal machine in and screwed it to the kitchen floor, right in the way between the cupboard and the stove. The room, was already too small for the people and the activities that took place. A handle hung at the side of the separator, which had to be turned at just the right speed to separate the milk from the cream. The discs through which the milk passed fitted into each other, and if put in improperly, the milk would not separate.

"I hate washing these discs," Ruth complained, "they make the dishwater all jelly."

Kate groaned. "Get clean water, and don't use soap. That's what makes it jelly. Be sure the water's hot or you'll never get them clean, and be sure and scald each piece after. If you don't, the milk will sour quicker in this heat."

The morning wore on. All other household tasks had to be done to make room for the washing. Heating the water on the stove made the rooms like an oven.

"I wish Bruno was here. He made everything clean so fast." Mary swept the floor, pushing boots and clothing ahead of her.

"You pick up those things and hang them properly — and put the boots in the bootbox where they belong, and never mind Bruno!" Kate looked at the mountain of dirty clothes and the hot sunlight outside, unrelieved by any shade trees, and sighed.

"I wish he was here too," she added.

Sam rushed in for his eleven o'clock lunch. He gulped the scalding tea poured from the pot on the back of the stove, and wolfed a couple of hot biscuits fresh from the oven. He sat down and propped himself against the north wall of the shack and closed his eyes.

The children whined listlessly about their mother's feet and Sam snored, oblivious to the activity and noises about him.

Kate piled the overalls in one heap and the flourbag sheets and pillowcases in another. She put homemade soap mixed with water on to boil, then added it to the copper boiler of soft water. The stove roared. The kitchen became unbearable.

The washing tub was yanked over the doorstep into the crowded kitchen, and filled with pails of hot water dipped from the boiler on the stove. The whitest clothes were dumped into the tub and the rocker fitted to the sides of the tub. It was shaped like a slatted potato masher and moved back and forth over the clothes, beating out the dirt as it rolled. Kate stood beside the tub and moved the handle back and forth.

By the time the washing was well started, Sam was up and away to work. He hitched up a four-horse team to a set of harrows and moved into the field next to the house to summer-fallow. Kate came out to hang the first lot of wash on the clothesline. The sun beat down with a blinding glare, reflecting on the two ribbons of steel that ran by to the south.

A sense of foreboding assailed her.

"Whatever is the matter with me? Is something dreadful going to happen?" She stopped the rocker and checked on the whereabouts of each child, and looked with anxiety towards the field where Sam was working.

The sheets and pillowcases flapped blythly on the line, a dazzling white in the hot sun.

"Anyway, that's a nice clean line full." She was pleased with her work and waved to Sam as he rounded the fence corner and came by the house. Clouds of lose black soil stirred up from the harrows changed direction and blew over the house, covering the line of washing with grey prairie dirt.

"Drat the luck! I'm not going to put those things through again. Maybe the wind will blow the dirt out!"

Kate worked on, but things seemed to drag and the pile of unwashed garments seemed no smaller.

Then the phone shrilled, one long, continunous ring. She dropped a wad of wet clothes back in the basket and rushed in to answer it.

"That's an emergency ring. Keep quiet, everyone!"

"War?" The anxiety in her voice attracted the children and they all pressed around her. "War declared? France — with Germmany? How awful! Britain too!"

She replaced the receiver in its cradle and sat down abruptly. The children plied her with questions.

"We're going to help Britain! Germany's going to fight France! Just wait until Bruno hears about it! He'll fix those Frenchmen! We won't have to worry if he's around."

Kate looked sick.

"Britain has declared war AGAINST Germany. We will be fighting WITH the French."

The children quietened. Mary and Ruth began to cry.

That afternoon a surprise visit helped cheer Kate. The sound of buggy wheels in the yard and a "Halloo" announced her sister Jessie's return. In a short time the washing was whisked away, dinner was cooked and the children washed and fed. The sisters had time to sit and discuss the war over teacups.

"You're working far too hard, Kate. You know your operation was a miracle. You're the first one who ever survived one around here. You should have help."

The crowded rooms now held seven children and two adults.

"You should have help all the time."

The declaration of war had a galvantic effect on the west. Harvest was just beginning and prices promised to soar. War was discussed in sober tones, but an underlaying excitement and optimism coloured all conversations. Young men who blushed and stammered in the parlour, suddenly lay down their pitchforks with their shyness. They donned the khaki uniform of the king and picked up their rifles along with the sure confidence of maturity. Waves of patriotism, like wind across the unharvested wheat, washed over the western plains. So many men flocked to the colours that the future looked bleak for farm help.

Sam was openly jubilant about the promise of prosperity.

"We'll get machinery to put in crops next year, and more to harvest it — if the war's still on — get rid of some of the stock — the horses —" Sam planned a year in advance in anything he did.

His ideas on a future of mechanized farming were no sooner announced than the answer seemed to appear with Markle's car. It chuffed into the yard with two official-looking men in the back.

After greeting them, Sam ran back into the house.

"Put on the tea and make some hot biscuits, will you, Kate? They want to buy two or three of the horses. Markle says I'll get top price for them, too!"

He trotted to the barn where the visitors were already examining Jim, a long-legged black yearling and the children's pet. He bustled in to direct their attention toward a heavy pair of work-horses.

"That one's no good — too frisky for work — he's going to be the kid's school horse. This team over here is the best, and I've another pair of Percherons out in the pasture —"

The men paid no attention, but continued to examine Jim and cast critical eyes on Meg, the young black mare who had foaled him.

At the house, the biscuits, golden brown, were just coming out of the oven, and the tea was put on to steep. The sudden rattle-bang-chug of Markle's car startled Kate. She looked out of the window as the car wheeled out of the yard and disappeared over the hill. Sam, hands jammed in his pockets, was coming slowly up the path.

"What's wrong, Sam? Didn't they like our horses?"

"Wanted them all right —" He slumped in a chair and looked out across the fields. "Offered any price for Meg and the black colt— I nealy hit one fellow!" He stared at the settling dust in the wake of the disapearing car.

Kate stopped in disbelief. Sam never argued with anyone, let alone threatened to punch them.

"Wanted them for the cavalry — no horse of mine is going into battle to be shot—"

The new prosperity and excitement of war turned sour when the grim facts of life and death and things to come confronted them.

War caused other troubles on the farm. Help was hard to get. Threshing dragged, lengthening the job and increasing Kate's workload. Jessie's visits were fewer and Kate ached for a change, any change, to break the monotony. Grandpa and his bride had made no effort to contact the family since their first disastrous visit. On a Sunday morning Kate decided to pay them a call.

"We'll be in town in about an hour, Father. Tell Elizabeth not to go to any trouble. What? No, just the two youngest. Yes, Well, good-bye for now." She jammed the receiver back on the hook.

"Hhmmp! Doesn't seem too anxious to see us. Funny thing. First time he ever mentioned not bringing all the children."

She thumped a chair into place and paused, then hurried blindly into preparations for the trip to town.

"Just needs some company, I'll be bound. Anyway, we're going. I want to hear about the conditions in the old country."

Mary and Ruth stood on overturned apple boxes and washed the breakfast dishes. Willie dragged in two sticks of green firewood and banged them into an empty woodbox by the stove.

"Mary, keep those children away from the barn. Don't let Buster go near my young goslings. Give the hens some water about noon, and don't forget the turkeys."

She struggled into her only starched white blouse. A yellow scorch mark from the ironing, forgotten since last washday, showed raw and ugly on the front. Long greyish-black lines of dust that constantly sifted through the windows lined each sleeve from collar to cuff. She ripped off the blouse and shook it desperately. Remembering the fashionable bride's visit, she sighed.

"She's had Father to contend with, and the housekeeping. Maybe she won't be so well-turned-out now," and Kate put the blouse back on again. She pulled on the only skirt left that would fit her now, and noted with desperation that it barely met over her middle.

"We may not be back until late, so see the children get their breadfood and put them to bed early."

She skewered a plate-shaped black straw hat, alive with waving ostrich plumes, on her piled-up hair, and a too-small black silk coat was yanked over her blouse and skirt. She had all of Grandma's clothes, but the only thing that fitted her now was the hat.

She climbed into the buggy. Mary handed up the baby, dressed in a white poplin coat and bonnet, a hand-down from the last three babies who had worn it. The next child, in a long-waisted white embroidered dress, summer coat, and shoes and stockings on usually bare feet, squirmed impatiently, standing up in the buggy between her father's legs.

"Watch the children, now. Don't waste the last of my brown sugar making candy, mind." The horses wheeled about and bounded out of the gate.

"She's a queer one, I been thinkin', Kate!" Sam slowed the horses to a steady trot and guided their heads across the rough stubble, a field unfenced and recently threshed, on a short-cut to Langbank.

"Don't know how she will take us. Doesn't want to see anyone, according to Father. We won't take offence if she's not 'ospitable!" He looked skeptical. "She's just lonely, I'm sure. Never married before, too, more's the pity! She just needs someone to talk to, maybe."

Kate set her jaw in her characteristic expression. This was the first trip she had had out in months and she was determined to establish a family solidarity with this attractive woman.

"She'll surely have some interesting things to say about London once we break the ice. Maybe she'll have some good books she can lend me."

An hour later the buggy was on the way home. The two children were crying. Kate was silent and close to tears herself.

Earlier they had arrived at Grandpa's. Kate, with the two children in her arms, knocked at the door while Sam was stabling the team. The old man opened the door. He had a broom in his hands and his moustache shook with rage. The room behind him, always so immaculate when his first wife was living, was in chaos. Trunks and partially packed boxes were piled everywhere. Broken dishes were scattered about the room and food was spattered on walls and furniture.

"Sam better take 'ee 'ome, Kate, The wife, 'er be off 'er 'ead."

CHAPTER 24

A few days later Grandpa Marks put his ailing wife in the Moosomin hospital and drove out to the farm for his first visit in many months. The Winnipeg daily newspapers, the family source for all world news, were churning out war information with little assurance that the stories were more fact than fiction. Grandpa's favourite brother, Captain Richard Marks, late of the Royal Navy, wrote often and gave news of relatives, Cornishmen all, who were now in sea service. Grandpa, as a true source of information, was pleased to relay the "facts" concerning the war, and also he wished to tell the family the medical report on his unhappy bride.

He was deaf, and yelled as if everyone else was hard-of-hearing too.

"Doctor zez take 'er back to the old country, else I must put 'er in Brandon Mental 'ospital." Dreadful stories, enlarged in the telling, came out about the "madhouse" in Brandon and the condition of its inmates, from time to time.

Kate stopped working and stared at her father-in-law.

"Father! You can't be thinking of committing the poor woman to that place!" Then she realized the alternative.

"But you can't cross the ocean with the war on. The Atlantic is swarming with German submarines! You'll be sunk!"

The old man's wattles trembled. His face turned red as a turkey gobbler's and his voice rose two octaves above his customary shout.

"They Germans! Kitchener will 'av un all sunk come Christmas. Dick will be called back to zervice! 'ee says war'll be over 'andily by then!"

He applied to the British War Office for permission to cross the Atlantic; backed up by his Navy Captain brother he received it, and left for England at once with his ailing wife. She lived for only a few weeks and died of a brain tumour before he returned to Canada.

This trip was the shortest on record. He returned on the Monmouth, on which his two nephews served, and took the train straight home. Kate and Sam, with all the family, drove into town to meet him and get him settled in his own house. He was silent on his wife's illness and death, but talkatively jubilant over the war and his adventures in London.

The children, ignored for months because of his wife's health, rushed about the house, helping him clear out the dust and happily inquisitive about his return luggage. Trunks and boxes belonging to the sick lady, which had been left behind, were pushed to one side to make room for the piles of books and toys he had brought home for each child.

As he unpacked, the table and chairs were covered with hats. He loved hats, and he came home loaded. A christie stiff, a rounded crowned hat with stiff brim worn by the British business men was still showing up at summer picnics or at church on the prairies. There were deerstalkers, plaid tams, a black opera hat which folded into a flat pancake and snapped up as if by magic into a stovepipe affair. He had several navy caps his naval officer brother had given him, and the prize of all — a German war helmet.

"I be walkin' over Lunnon bridge with this 'ere 'at of brother Dick's on. and it be dreadful foggy. Two policemen stopped and took I to Lunnon Tower. I be questioned all night! Thot I be a German spy!" He roared with laughter, told the story repeatedly, and relished the chance to tell it to all who came to the house.

Later the children sat about quietly reading their new books and munching warm pasties. Kate dawdled over cups of strong tea and encouraged the old man to tell of his adventures overseas. Sam went off to the Post Office to collect the mail and the newspapers. He came into the house quietly and handed his father the Winnipeg Free Press. Black headlines covered the front page.

"MONMOUTH SUNK WITH ALL HANDS ON RETURN VOYAGE. NO SURVIVORS." The old man's face turned black with rage. He opened and closed his mouth but no words came for some moments. Then —

"They 'un! The murderin' 'un!" He sobbed and seemed not to notice his visitors for some time, pacing the floor in a kind of frenzy.

"No survivors," he muttered — "an' no survivors!"

For the duration of the war he glared, thumped his cane on the floor and muttered openly about "they 'un" when anyone suspected of having German blood crossed the threshold. He was not alone in his views.

Many were suspect. The war propaganda steam-rollered across the plains. Page-high pictures of King George V, Kitchener, or a young, handsome soldier beckoned from every newspaper. One advertisement using a giant ear and the caption, "WALLS HAVE EARS", below it lent credence to the suspicion that the prairies were awash with disguised spies, all ready to cut throats or open gates to armies of goose-stepping enemies.

Everyone became an amateur Sherlock Holmes.

A Red Cross sale was held in town to raise money for comforts for the soldiers. Every woman must contribute something, including a Mrs. Brown and a Mrs. Smith. Kate had a gossipy visitor the next week.

"Did you note that cake of Mrs. Smith's? Icing two inches high! Maybe she hasn't heard of sugar rationing! Just maybe! They came up from southern Ontario where all those German settlements are! I wonder —"

A Mrs. Brown had donated a plain cake for the sale — with no icing.

"Mmmm," remarked the woman when Mrs. Brown's name came up. "Not much interested in the war effort — her! Did you notice her cake? Looked heavy to me, and no icing. Keeps mum about where they hailed from. Wouldn't I like to have a look at some of her mail and see who's giving her orders!"

Family trees assumed importance.

"My Scottish grandmother", "my French father" , or "when my family moved out from Ireland" were casual but calculated interjections in every discussion, even among formerly close friends. No one must be left in doubt as to what nationality you were.

For months Kate and Miss Denker, the hospital matron who had helped save her life, had enjoyed a lively correspondence and had exchanged photos. Miss Denkar's picture was put away. Kate didn't answer any of the letters which came for a while and then stopped.

Into this hotbed of suspicion in the village, came the family of Dawes. They arrived in town with twin sons, overwhelmingly British accents, and the family cow. It was the cow that started the trouble.

Each family owned, and was responsible for, their own milker. The cattle were pastured a mile away and were brought home at night by the youngsters of the families who owned them.

Herbert Rempel was the town clown. A gangling youth, a head taller than the other children and miles ahead when it came to devising new ways of tormenting the Dawes. He never used violence, but his tricks kept the village in a constant uproar.

He slipped garter snakes down the twins' necks. He left early for his cow, chased the Dawes' milker to the back corner of the pasture and hid behind a granary where he waited until the twins came by in the dark. Then he came out at them, yelling and waving a hatchet, dressed in a gunny sack with feathers in his hair.

Every milking time was punctuated with the screams of the twins as they escaped from Herbert's latest torment and rushed for the protection of their mother's voluminous skirts. Mrs. Dawes shouted references to Herbert's supposed canine background or made definite declarations that Herbert's parents' marriage had never been sanctified by the church, and all on German soil to boot.

The louder the shouts, the happier Herbert was, and the merrier his whistle as he loped along home with his own cow.

The twins, in spite of their mother's over-zealousness, caught every germ afloat and were the first ones to get measles. The first day they were bedridden Mrs. Dawes, frantically trying to contact the doctor, wringing her hands and taking their temperatures, forgot all about the milk cow.

The doctor came, examined the boys, put the house under quarantine, and left. Night fell. Mrs. Dawes, frantic with worry, opened the door to a thumping knock.

She was confronted by the smiling face of the hated Herbert, his straw hat in one hand and a full pail of milk in the other.

"Good evening, Mrs. Dawes!" remarked the tormentor, handing her the pail. "The little German bastard has brung home and milked yur cow for yuh."

With an old-world flourish he clapped his hat on the back of his head and was off down the lane, singing "It's a Long Way to Tipperary".

"Tom Cairns away to sign up this morning. I hear Ervine Johnston and Bob Young have gone too. Upton says they'll be at the front before Christmas. Top men, every one of them! What a waste!" and Sam handed Kate some letters.

"Well, they're young. It's their duty. The country needs them and it's their job to go." She slit the envelope open and began to read. She read for a few moments in silence, then she put down the letter and fumbled for her handkerchief.

"Sam! It's from Donald — my brother —"

"What's the matter, Kate? What's happened?" Sam grabbed for his wife, as she looked ready to faint.

"Sam! My brothers! Both signed up. Off to war! I can't believe it. Why did THEY have to go? It's not fair!" She began to cry and disappeared into the bedroom for the rest of the day. Sam made the supper.

From then on the world seemed out of joint for Kate. A kind of desperation set in, and events which seemed out of time and out of step took place.

Her sister Jessie, finished with a nursing case, came home to see if she could help and hurry the building of a new house the family needed.

"You MUST take things easier, Kate," she worried. "Your life depends on it." Kate looked pale and sick.

"I'm all right, Jessie. Would you leave me alone!"

"You should be in hospital right now. Two and a half months to go and all this work! You won't make it unless you take things easier!"

"I'm not going to any hospital. I've had all the hospitals I want." Kate ended the argument by slamming the bedroom door after her. Her sister gathered the children and went for a walk.

They came back shortly.

"Kate! Kate! We've found an ideal place to build the house. A little hill to the north of here, and a clear spring right by it. Good water, and it's soft, too." Jessie was jubilant, but Kate still sulked.

"I don't care where it's going to be built — if is ever gets built." Kate was bitter about the moves she had made already, from places of beauty to this bald, ugly spot, and the enormity of the task was too much for her.

Sam came home from town the next morning with more news. His father, happily expansive, sent word to Kate that she was to have the new house in Langbank, and it was to be moved whenever she settled on a spot to move it to. With the gift the old man came, too. Sam was jubilant. Now he needn't build.

Kate was less than enthusiastic.

"I am NOT looking a gift horse in the mouth! And I won't have him around shouting. I won't put up with it. You tell him, or I won't move a foot out of here. Besides, where can we put it on this Godforsaken bare plain? Not a tree in sight, or a pretty spot to set it!"

Grandpa Marks, a head shorter than any member of the family, made up for it by acting the martinet. His forceful personality, fierce manner, and ability to be heard across a section of land on a windy day, helped him maintain this authoritative position. His sons never challenged him. The old man was pleased with himself, and waited for Sam, who would be loaded with gratitude, to return. Instead, Sam arrived looking glum.

"Father, Missus is nervy — can't stand noise — your yelling — wants me to tell you ——" he apoligized lamely.

Grandpa stood still for several seconds. His eyes became suspiciously moist, and his moustaches trembled. He took out his handkerchief and blew his nose. Then he said quietly.

"I not be yellin', Sam." He fumbled for a stub of a pipe, lit up, and turned his back on his son. When he turned around again, he assumed his old fierce front, and at once began to put the house in order preparatory to the big move.

CHAPTER 25

Harvesting was slow. The war had taken all the able young men. A few middle-aged neighbours and young boys, some scarcely able to pitch a sheaf over their heads, were the only available help.

Sam finished his own threshing and pulled his outfit into a batchelor neighbour's place. The cooking fell to Kate. Jessie, bone-weary after a nursing job in which her patient had died, tried to shoulder some of her sister's burden. The engine broke down and the threshers milled about in the tiny house, getting in the way. Into this unhappy situation bustled Grandpa Marks.

He had packed everything breakable in his house, nailed down everything moveable, and presented himself at the farm to help with the move.

"Father, come to town with me. Machine's broke down and I have to get repairs. Have another engine to line up in Kennedy to pull the house. Threshing will be over tomorrow if I can get the repairs. We move the house the next day. Two carpenters from Wapella are starting on the barn. Have to get it up before the snow flies."

Sam shouted over his shoulder, busy scrubbing off chaff and tractor dust. He hurriedly ate some lunch, drank his tea standing, and was about to take off when his father stopped him.

"Sam! Sam! 'ee can't putt the 'ouse on the bald prairie. Must putt a zeller down first — cement — else the floors will freeze. 'ee can't putt the 'ouse on the bare ground!"

"We've no time to do it now. I have to get the engine before it snows. We can dig it out in the spring."

Jessie glared angrily while Kate, unequal to the work at hand, the move, and the clash of personalities, escaped to the bedroom.

"Get un fixed now, else it will cost more," objected to old man.

Sam scarcely paused to listen. He rushed out the door. His father, at his heels, grumbled and snorted with rage and frustration.

In town they bought repairs, lined up the engine, and were getting into the wagon to come home when they were stopped by a tall, angular man. He stepped up and blocked their way, stuck out his hand and rocked back on his heels.

"Well, Sam! How are you? Haven't seen you since that mare of yours won the race at Cannington." He put a boot on the wheel, took out a chew of tobacco, cut off a quid and offered the plug to Sam, who didn't chew. He ignored the old man. Sam hesitated, nodded and tried to remember the man's name.

"I made a few dollars on that race. You always were a top man with horses, Sam." The fellow seemed in no hurry. Gandpa Marks muttered behind him and tried to get into the wagon.

"Hear you're movin' a house. That Titan of mine is just sittin' in the yard doin' nothin'. When's your move?"

"Day after tomorrow. But thanks, I've half a day's threshing to finish, and I've got four engines now. What did you say your name was?"

"McGuire. Don't you remember?" and he gave Sam a friendly nudge, still barring the way.

Grandpa Marks, Cornish cautious, reared back and muttered darkly at the instrusion.

"Day after tomorrow, eh? Don't say another word, Sam. You done me favours with that horse of yours. I know all about these moves and you can always use another friend. I'll be there to help or my name's not Andy

McGuire." With a wave of his hand, he left before any further objections could be made.

"Oo be 'ee, Sam? 'ee be might suspicious to me. Don't trust un, I say."

But Sam, whipping up the horses, was intent upon getting back to the threshing, and had forgotten the man already.

On moving day the house was jacked up and put on skids. It was to be dragged two and three-quarter miles across the prairie, east to where it would be set up on the chosen hill.

McGuire arrived with his engine just as the others were about to start the pull. He wheeled it into place before Sam had a chance to speak to him, and was allowed to continue. Before the house budged, McGuire's engine sputtered and died. He unhooked it and tried to start again, to no avail.

Grandpa Marks stood back and as the house ripped slowly from its foundation, burst into tears.

The other men pulled away and trundled the house slowly across the fields to the east. Hayracks and wagon loads of furniture followed. By afternoon the house was in place on the hill, and the sides shored up with a few stones on bare ground.

The furniture from the little house had been moved too, and supper was on the table. Jessie put the children to bed. Grandpa Marks shuffled back and forth, still growling and angry about the lack of a cement foundation. He made pots of scalding tea and cooked mounds of meat and potatoes, and served them to the tired men. Kate sat at the table, white and spent, listening to the account of the big pull.

" 'Ee ain't 'eerd the last of that McGuire," prophesied the old man. " 'Ee be a crook as I've 'eerd tell." He stuffed more wood in the fire and pulled more biscuits from the oven. He was stirring up the fire with the poker when the subject of his predictions rumbled into the yard and tied up his team.

McGuire strode into the house, hardly taking time to knock, and came straight to the point.

"Sam, I'll have my $65.00 right now. My engine was broken by a loose two-by-four from your house. $65.00 will just barely cover it." He stuck out his hand to Sam, his unpleasant face grim and threatening.

The men around the table stared open-mouthed at the intruder.

"I don't remember hiring you in the first place," Sam answered mildly from his place at the head of the table. "Your engine did'nt take part in the pull, anyway. You quit before we started."

McGuire lumbered further into the room. He towered over Sam, shook his fist and snarled, "I want that $65.00, and I want it now!"

For a moment there was a stunned silence. Suddenly Kate jumped from her chair at the end of the table, grabbed the poker from Grandpa Marks' hand, and charged the burly giant who threatened her home. All the surpressed fury of the past weeks of work, the inadequate equipment and the mounting family problems to which there was no satisfactory answer, welled up in her and burst forth. She swung wildly at the bully and caught him across the face.

"You scoundrel," she raged. "How dare you come into my house. You're a crook and everybody in the country knows it. GET OUT! GET OUT THIS MINUTE!"

She swung again and slashed him across the arm as his hands flew up to protect his head.

"You scoundrel! she shouted, and then the ultimate insult — "YOU GERMAN SCOUNDREL!"

The attack was so sudden, not a man had time to move before McGuire, a red welt across his face and with a bloodied nose, was running for the door, arms about his head in abject retreat.

The whole table heaved at once. Chairs went flying and the men, in a body, rushed to protect Kate. She needed no help.

McGuire sprinted for the safety of the wagon box as all the dogs, aroused by the shouts, leapt up in pursuit, snarling and barking at his heels. He loosed the reins and slashed the whip to the horses. The men behind him roared with laughter and slapped their overalls in delight.

"You'll hear from me, Sam Marks!" McGuire shook his fist at the group who had watched his defeat. "You'll hear from me yet!" His horses, terrified by the hubbub, sprang to the gate and were off down the road in a thundering gallop.

Kate marched back into the house, head up, hair falling around her ears, and her eyes bright. She replaced the poker on top of the stove. Jessie rushed into the room from upstairs, where from a window she had witnessed the end of the altercation.

"Oh, Kate! How could you? How could you?" she exclaimed.

Kate looked sheepish for a moment, then began to laugh. Everybody joined in, even her sister at last. The day ended in weariness but good spirits. Beds were soon filled, and in a short time the house slept.

Day was just beginning to break when Sam shook Jessie awake.

"Wake up, girl! Kate wants you. Her time has come, I think."

Jessie shot up out of bed and grabbed for her dressing gown.

"It can't be Sam. She still has some weeks to go. Go for the doctor, hurry!" Jessie rushed into the bedroom to find her sister in advanced stages of giving birth.

"If you don't get the doctor, I won't be responsible for her life. Hurry, for God's sake!" Jessie, teeth chattering and hands shaking, prepared for the worst.

The doctor arrived hours later. By that time a tiny morsel of humanity, a four-and-a-half pound baby girl, had arrived. Jessie was feeding her with an eyedropper. The baby's bed was a tiny box lined with cotton batting. The usual loud bellow associated with Marks' births was missing. Only a soft mewing sound came from the baby as the children crowded around to see the new sister.

Mike Hayes, an affable Irishman and the new hired man, sat after supper surrounded by the children and puffed at his pipe.

"Did you see the new baby, Mike. It's so small and it doesn't even cry. It just mews like a cat."

Mike pulled at his pipe. "Well, I guess we'll just have to call it Kitty then, won't we."

McGuire sued Sam and lost. He appealed the case, hired a crack lawyer and won, and Sam was forced to pay for the cost of the court plus all the attendant expenses. There was no extra money to spend on a cement basement or many other things needed around the house. The problem of a house with no storage space was accepted and forgotten. The cement basement never did become a reality.

CHAPTER 26

The winter following the move was long and bitterly cold. The exposed knoll on which the house sat caught the full force of the driving winds which poured out of the north. No basement, no furnace and inadequate banking made it unbearably cold. Moving the house had shaken loose and unputtied the windows. The resulting drafts licked up the feeble heat from the stoves. Spilled water turned to ice on the floor during the day. Any water or food left in containers froze solid by morning.

Kate battled to keep the premature baby alive. Housework was neglected, with chaotic results.

"Boys, fill the woodbox now. Bring in some dry wood along with the green stuff, for kindling in the morning." She hoped jobs would get done without having to check them.

"Grandpa, will you fire up so the baby will be warm for her bath? And I'll need a boiler of snow, too, for the bathwater. The well water is so hard."

Breakfast dishes littered the table. Beds remained unmade and floors unswept while she fed the baby.

She opened the oven door and put a large wash bowl of warm melted snow water on it.

"Bring the baby's soap, will you, girls, and get at those dishes." She sat on a rocking chair by the stove and carefully unwrapped the cocoon-like swaddling clothes about the tiny girl. Work stopped and the children gathered around to watch the ceremony of the bath.

"Stand back, now." Kate lined the bowl with a clean flour sack. The tiny head was wiped and patted dry. The baby shivered and the other children listened in wonder to the feeble objections coming from the naked child.

The kitchen door burst open and great clouds of frosty air swirled into the room.

"Shut that door!" Kate threw a towel about the infant. Buster pushed in with an armload of ice-encrusted wood and banged it into the woodbox by the stove. A cloud of snow rose above it.

"Keep back! Keep back!" shouted Kate.

Willie followed with his armload of wood and the kitchen temperature dropped to intense chill. The two boys, still in frozen mitts and snow-covered coats, pushed into the circle. Kate vainly tried to protect the shivering mite with towel and blanket.

"Keep back from the baby with your cold clothes. Can't you see she's having a bath?"

The fire was re-stoked, the warm water renewed in the bowl, and all cold clothing removed from the vicinity before the bath again began.

Kate carefully washed and dressed the blood-encrusted umbilical cord.

"What's that string on its stomach for?" The ring of faces squirmed in closer.

"That's where it got its food before it was born and was attached to its mother." Nobody questioned further. Birth, like death, was accepted as part of other farm realities.

"Doesn't it hurt the baby?"

"No. It's part of nature and will soon dry up and heal." She wound a wide band of cotton around the baby's middle and pinned it in place. A diaper was pinned on next.

"Hand me the barrycoat from the warming oven." A band attached to a long skirt was wrapped around the child. The skirt was folded over the tiny legs and tucked in. A shirt and another long skirted slip was put on followed by a long flannelette dress. The baby was wrapped in a blanket and placed in a makeshift box bed. The box was placed in a black wooden cradle, stuffed about with more pillows and blankets and set near the stove away from the blasts from the constantly opening kitchen door.

"Mary, punch down the bread dough, will you. Ruth, go upstairs and make the beds and empty the slops. Hazel, you dry the dishes for your grandfather."

She rose, picked up the scattered clothing, surveyed the dishevelled room, and sighed. Mary was pounding the stretchy dough with tiny hands, and trying to lift it back into the pan. It had overflowed and was now stuck to the towels covering it. Grandpa Marks, pipe in mouth, was slopping the dishes through a pan of cooling water. Some of the children played noisily under the table. Baby Amy sat in the middle of the floor and cooed happily at a broken doll she had bedded down in an old overshoe.

"This house is no easier to keep clean than the last one," Kate muttered. She stirred up the fire and began to take the separator apart for washing.

Winter dragged on. The baby thrived. Kate didn't. A queer lassitude settled on her spirit which even the first welcome cawing of a crow failed to lift. Unexplained pain knifed through her periodically, warning of trouble to come.

One morning Sam found her slumped over the end of the bed, half-dressed and unable to speak. Within hours the two youngest children were bundled off to a neighbour and Kate was lying, wrapped in blankets, in a straw-filled sleigh box on the way to the hospital thirty-five miles away.

Her arrival at the hospital was dramatic. The doctor who had operated on her the previous year had closed his office and was preparing to leave for the war front. He strode into the hospital, supposedly for the last time, dressed in army uniform, and heard Kate, half-delirious, calling his name. He took off his uniform, donned a white smock, and began to scrub up for the operation.

'Ach, Mrs. Marks! Too soon! Too much vork!" The kindly round face of Miss Denkar bent anxiously over the stretcher being wheeled into the operating room.

"Too soon! Too soon! Ve fix!" Within minutes Kate's second major operation in two years was in progress. A mended hernia and six inches of bowel removed when the operation ended. The worried staff watched anxiously by her bedside.

For several hours consciousness came and went. Kate couldn't open her eyes. When she did, it was to see Miss Denkar's worried face bending over her.

"Forgive me! Forgive me!" Kate muttered and fell asleep again. Hours later she wakened to find Miss Denkar working quietly around her.

"I'm so sorry. I didn't write. My brothers — gone to war! I thought — I hated everything German — not your fault —"

The nurse put down the thermometer and the hot water bottle she carried, sat down by the bed and took Kate's hands in her own.

"Ach! Ach! Mine brothers, too. Both dead — in German army." Her efficient composure crumbled and a torrent of tears, like a broken dam, mingled with Kate's in their common grief.

CHAPTER 27

Sam returned home in several days with good news.

"Kate's on the mend, Father. I'm going to Kipling tomorrow to see if I can get help there. Kate'll need it when she gets home." He left his father to cope with the household.

The next day the house was in chaos. Books, boots and discarded clothing lay everywhere about the floors. Grandpa Marks walked back and forth from kitchen to dining room, tracking through a pan of spilled flour. He had an egg-lifter in one hand, and a frying pan in the other. From them he dispensed fried eggs, browned potatoes, and hunks of fried bread. The seven children gobbled, clattered and fought around the cluttered table.

Mike, in from chores, ate his breakfast, lit his pipe, and watched activities. The two smallest, still in sleepers, knelt on a long bench by the table. They helped themselves to milk, cream and sugar, which they spread liberally over plates of porridge which overflowed on the table oilcloth and dribbled over the edge to the floor. A black dog, a collie, and two tiny white fox terriers yipped and scuttled about under the table, snapping at the food that was spilled or was thrown to them.

Amy had just learned to eat by herself. She shovelled porridge from plate to mouth, losing half on the way. The residue lodged in the creases of her sleepers and down her front. The back flap of her sleepers hung down and exposed a rosy backside like two loaves of fresh baked bread. She slid down from her place and was surrounded at once by the dogs.

"Eat here," she directed, pointing to her porridge-laden stomach. Mike laughed and lit his pipe. Amy, cleaned to her satisfaction by her grateful canine friends, backed up to Mike.

"Do up my tail-gate, will yu, Mike?" she requested.

Mary jumped up from the table as the clock bonged eight.

"I'll make the lunches, Ruth, if you find Hazel's clothes."

The two oldest girls rushed through their jobs. Sliced, crumbling bread was buttered and covered with a fried egg mixture and put into jam pails. Their long hair had to be combed and braided, and they worked with feverish haste.

"Grandpa! I can't get the brambles out of my hair!" Hazel clung to her Grandpa's overalls and held out a semi-toothless comb.

"Wullie! See 'ee get the team out and 'urry, else 'ee'll all be late."

The old man set the sizzling pan on the oilcloth table cover and took comb in hand to attack the tangled hair.

The rush of the dogs to the door, yelping and barking, and the sound of a wagon outside announced Sam's arrival with the promised help.

The children, ready to leave for school, pushed out of the door. Their Grandpa was right behind them, stump of pipe clenched in his teeth, and the broken comb still in his hand.

"Father, this is Alice. Got off the train this morning. Show her where she's to sleep."

"Alice? Alice?" Buster looked up at the newcomer with surprise and some disgust. He was accustomed to pretty farm girls who came, stayed for a few days, and left. "Why, she's an old lady. An OLD Alice."

"Old Alice", perched on the wagon seat beside Sam, with her clothes in a bundle over a short, stubby arm, looked like a round, brown bird. She wore a rough homespun garment, clinched in the middle by an unseen belt, like a string around a pillow. Her little white face was almost hidden by a large babuska tied under her chin. Two black shoebutton bright eyes, tucked in the folds of her face like currants in an unbaked bun, looked down on her future charges and twinkled.

She floated down from her seat over the wheel, light as blown thistledown. Her skirt swung and swayed and belled out around her, revealing layer after layer of snowy petticoats. Tiny slippered feet peeped out beneath, touched the ground, and moved swiftly towards the door.

Grandpa urged the children off to school, and the little lady entered the house and disappeared upstairs. Sam turned the team towards the barn and left his father to take care of Alice.

Minutes later she stormed down into the kitchen, sputtering invectives in a mixture of English and Hungarian. The twinkle and the smiles were gone from her eyes. She sparkled with anger as she piled wood into the stove and put the boiler with water on to heat.

By midmorning she billowed down the stairs backwards, dragging a big wooden scrub pail. The soapsuds flew and great sounds of grunting and disgust accompanied the activity. Sam rushed in for his morning tea and left just as hurriedly without it. His father smiled and sat contentedly peeling a huge pail of potatoes.

Sam and Mike hauled in rack-loads of straw from the winter stacks all morning. By noon, hungry and weary, they unharnessed their horses and went to the well for a drink.

"Smells like we will get something to eat," Sam remarked. "She looks like a bearcat but the Missus will be pleased if she gets the house in order."

As they approached the house the smell of Hungarian cookery, spicy stewing meat and rich raisin pie laced with strange flavours, was reassuring, and they went in to a clean, well-ordered house.

When the children returned from school, Alice took each in turn as they entered, and by gestures and broken English directed them to a row of nails she had driven in the wall behind the kitchen door.

"Heng! Heng!" she directed as coats, sweaters and caps were each put on their own nails for the next day. This was the first lesson in tidiness. After school, only a sharp look from Alice was needed as a reminder that clothes must be hung up all the time.

Before the plowshares bit into the windy grasslands of the prairies every knoll across the land was riddled with gopher holes. These prairie rodents scrambled out in the spring to whistle their defiance at the world. Knots of garter snakes used their holes to hibernate in the winter.

Gophers transferred their homes immediately when humans appeared. Snakes were never in such a hurry. When the first warm days of spring arrived they crawled out and sunned the winter aches from their long coils. The hill chosen for the house was one of these knolls.

One warm spring day the children arrived home from school early. Alice was busy with a great vat of yeasty dough. She had worked and stretched it the full length of the dining-room table, to the consistency of thin elastic webbing. From the dough she cut tiny, blintz-like morsels, onto which she put spoonfuls of hot fruit, a mixture of dates, raisins, currants and dried apricots which had simmered on the back of the stove. She sealed them into pockets and baked them slowly in the oven.

Her sleeves were rolled up to her elbows. Perspiration beaded her upper lip and forehead. From time to time she swung ropes of dough above her head as she stretched each section and worked it over before cutting it.

Grandpa watched the process with Buster, Edith and Amy hanging about the table top and watching with mouth-watering expectation. Alice fluttered and moved about like a ballerina centre stage, from table to stove to pantry shelf.

The rattle of the returning school buggy, as it stopped outside, broke the spell. "We're having pie envelopes for supper!" Buster banged out the door to make the joyous announcement. At that moment a monstrous black and yellow snake slid across the step in front of him.

Lunch pails went flying and Mary and Ruth screamed in alarm. The reptile, in no hurry to give up its usual sunning place, held its ground, curled up a bit and surveyed the interlopers with swaying head and darting forked tongue.

Alice, alarmed at the screams, dropped the dough she was stretching and rushed outside as the snake was about to move away. She hesitated a second, then grabbed it by the tail and swung it twice around her head. The children stood in open-mouthed amazement. She cracked the snake like a whip and the head flew off. She gave it a heave and it went sailing into the garden, its body quivering and still wiggling in protest. Alice wiped her hands on one of her petticoats, came in, scrubbed up, and went back to her baking.

When Kate returned from the hospital, she found to her amazement that the woodbox was almost always full, the school books and clothes were always put away with little argument, and everybody was early for meals. Anyone who had the courage to pick up a live snake by the tail and crack its head off was more than worthy of respect and obedience.

Alice left one morning with the same merry twinkle in her eyes she had had on arrival.

"Gee Wullie! I'll never forgit her. I'll shore miss them envelope pies of hers." Buster sniffed and lined up with the other children to bid her a sad goodbye.

Fifteen hundred trees were ordered from Indian Head Experimental Farm. Ash, Russian poplar, caragana, Balm of Gilead and prairie willow were planted around the house yard acre. They grew rapidly under Grandpa Marks' constant care. Soon Kate looked out in summer to a wall of green on all sides. She had her beloved trees at last.

CHAPTER 28

The six lines of young trees from the Indian Head Experimental Station needed constant care. Grandpa Marks hoed from morning until night. It took him a week to reach the place where he had started. By that time, the weeds were up again and racing over the rich prairie soil.

Amy and Kitt followed along behind him in bare feet, playing in the fresh-hoed lanes of black soil. They paused when he paused, and asked questions. The old man stopped to light his pipe and survey the flat plains stretching to infinity on all sides.

"Do you call this little weed pigweed because pigs like to eat it, Grandpa?"

"Whudd'ee say now? Whudd'ee say?" He puffed to get his pipe going and bent again to his hoeing, wrapped in his own thoughts. From time to time he stopped to pick the tender green weeds. He called the children to drag up a big dishpan to put them in.

"Do pigs like it like we do when you make boiled greens for dinner, Grandpa?"

"Eh? Whudd'ee say?" The questions, seldom answered, continued as the sun climbed up the spring morning sky.

"Father, could you help me fence this afternoon?" Sam drove up with a wagon-load of wire fencing. "I want to put it around these trees."

The old man stopped for a moment in his hoeing.

"I ben thinkin', Sam. War's on now and they be gettin' good prices more than likely for pork." He relit his pipe.

"Whyn't 'ee buy some good pigs. Make a pasture to the west there. Willie be big enough to 'elp of un now."

"Good idea, Father. Dean's got some razorback sows ready to farrow. I'll see if he has any for sale.

They bought extra wire, fenced the houseyard and a pig pasture. A narrow runway connected the pasture to the pigpen beside the well. The pasture was sown with rapeseed.

When the rapeseed reached a few inches growth, some dozen fat sows, ears flopping and udders swinging, each with her complement of squealing, protesting young, were dumped into the pasture. They wobbled continually back and forth all day long, filling the air with their nerve-grating sounds and nauseous odours. At night they settled in their pen and burrowed in heaps of straw laced with mud to keep warm.

"We're sure surrounded by them brutes," Willie viewed his new job with some distaste, "but I guess I can stand the smell. Dad's giving me one for my own."

With Grandpa's help the house became more orderly. The general feeling of prosperity from prices rising on pigs and other farm produce, as a direct result of the war, had an exhilerating effect on Kate.

"I'm going to raise ducks this summer. We'll have roast duck for Thanksgiving and Christmas. I got a clutch of eggs, a special kind, from Mrs. Cairns. They can swim in the slough behind the barn."

Kate bustled about, eager to get her project started. Grandpa Marks built a coop for the broody hen who would do the hatching. The coop was placed by the fence in the shade of a leafy shrub for coolness, and Kate watched over her experiment with the ardor of a dedicated scientist. She saved the best table bits for the hen, and checked her several times a day.

One morning she paused and leaned out of the upstairs window. For miles on every side, the black fields were spiked with tufts of emerald. The meadowlarks, singing from fenceposts back of the barn, filled the air with their joyous trilling. Perfumed Balm of Gilead trees scented the wind.

"It won't be bad here when the trees are full grown." She sniffed the fresh breeze, which for once didn't smell of pigs.

"Mama! Mama!" A scream of alarm from below jarred her into action. "Come quick! Your hen and a pig are having a fight!"

Kate raced down the stairs and out to her precious embryo ducklings. The coop, so carefully watched over for two weeks, was tipped on its side. The frantic hen, wings buffeting and head angrily jabbing, was flying repeatedly, with little effect, at a great white razorback. The nest of eggs was crushed. Half-formed ducklings lay trampled under the feet of the sow as she gobbled up eggshells and all that remained from the precious nest.

"You devil! You she-devil!" Kate picked up a stick of firewood and slashed the sow across its egg-spattered snout. It shook its head and waddled off towards the break made in the new fence.

Sam and his father drove into the yard in time to see Kate, with loose hair flying, a block of wood in her hand and murder in her eye, take after the sow. She staggered up the hill after the unequal combat.

"Just give me a gun! Give me the pleasure of killing that brute, will you. The devil never was driven out of those accursed animals!"

The two men both burst out laughing.

It was the wrong thing to do.

"Just you get that fence fixed, or I'll kill every hog that comes in the yard," and she disappeared upstairs. Sam, with his father's help, made the dinner.

That afternoon Grandpa Marks fixed the fence, built a pig-proof hen coop and brought a new setting of duck eggs from town.

It just wasn't the same. When the ducklings hatched some weeks later, Kate regarded them listlessly. It wasn't her own dream any more.

CHAPTER 29

Nobody at the Marks' house ever heard of Pavlov's dog, but all had a parallel experiment with Willie and his pigs.

"Willie! Get up, son — it's near seven."

Sam took a thin board and made curl slivers on it and placed it in the firebox of the stove. He piled dry sticks on top and one or two small green ones, lit the slivers and put the lid back on the stove. He dipped water from a pail, filled the kettle and put on a pot for the porridge.

Willie, sleep still glueing his eyes, rolled out of bed in the room above the kitchen and jumped into his overalls. He stumbled down the back stairs, clothes in hand, to finish dressing by the crackling cook stove. His leather boots were stone hard and covered with dried pig manure and prairie mud. He paused to dash water on his face from the wash basin.

"Hurry, son! Don't forget to close the chop bin after you feed the pigs. Hurry now." Sam rushed off to the barn and left Willie to follow.

The boy picked up a bucket almost half his own size and set off on his round of chores. He half-filled the pail with chop (ground barley and oats) from the bin and turned towards the pig-pen. As he went he tapped the side of the pail with a stick and began to whistle his favourite tune.

The moment the first notes of "Casey Jones" split the early morning quiet, the blanket of straw in the pigpen began to heave and move. The sows reared up, scattering their round pink piglets in a tangled, squealing mass. They wallowed to the trough grunting their pleasure while Willie, still whistling, began to dump in their morning mash.

"Whatever is wrong with those accursed pigs?" Kate, used to the constant uproar made by some three hundred grunting animals, became, aware, one hot fall day, of more than the usual annoyance of sounds. She noted a constant regularity of the screeching herd as they thundered back and forth at the foot of the garden.

The democrat, piled with children, books and empty lard pail lunch-buckets, clunked into the yard. The girls jumped down and came in to relate the day's happenings at school. Later Willie burst into the kitchen. "Mama! Mama! You gotta do something to Buster. He's spoilin' my pigs!" Red-faced with anger, he yanked at his mother's skirts.

'What do you mean, spoiling your pigs?"

"You just come with me and listen to what he's doing!"

They went outside.

"You listen to him! He's upsetting them after all my training!"

The pigs were bunched at the foot of the hill near the trough, squealing and shoving. Buster, with his back to the house, was whistling, a faint shrill above the din. Kitt and Amy stood watching close by.

When the noise was so great at the empty trough that nothing but squeals could be heard, Buster dodged along the fence half way along the runway to the pasture, followed by the two little girls.

He began to whistle "Casey Jones" as loudly as he could, while he tapped on a tin can. The pigs plunged and crowded about until the sounds on the pail became audible, then the whole crowd wheeled, snorting and squealing and tumbling into each other. They rushed along the fence towards their expected treat. When they were well piled up on the other side of the fence, Buster turned and ran back to the well and began whistling and pounding again. The pigs followed him back along the fence, their deafening noise filling the air.

"I've trained them myself, Mama," protested her eldest son. "I bang the pail and whistle "Casey Jones" and then they know I'm gonna get their feed and water. Now he's spoiled them almost. He's been doing that all day and tomorrow they won't come for me."

Kate tried not to laugh. "You go and have some milk and some fresh gingerbread. I'll attend to our Pied Piper." She hurried towards the screeching, disappointed pigs.

She reached the trough and was right behind Buster before he knew it. He was puckering up for another whistle.

"Well, young man! I see you know how to help with the pigs. You just pump that trough full of water, and you can carry all the pig feed tonight for Willie. You can also get up every morning and help him, seeing you are so good at calling them. GET BUSY!" She thwacked her enterprising son across the bottom with a switch she had picked up on the way. Buster, perspiring from an afternoon of playing this exciting game, was not ready to pump water for the horde.

'Gee, Mama, I was just experimentin'. I got to fill the woodbox. They're Willie's pigs!"

"For a week you'll do both jobs, and if I hear any more of your experiments, You'll keep them up for always!" She tapped him again on the bottom with the switch, watched while he got busy at his job, and went back to the house hardly able to hide a smile.

CHAPTER 30

Frost came early that fall. Threshing was over. A windstorm blew out of the Arctic, rattling the windows and snatching the heat from the chimneys.

"Sam, for heaven's sake, get some banking around the house. The place is freezing on our feet. We can't wait until the snow comes to bank it. There may not be enough for a long time yet."

Kate was beginning her ninth pregnancy and felt sick and short-tempered.

Sam was innovative if he could find a short cut and do two jobs at once. He went out with the men to get some banking material and passed the outside pigpen, now empty since shipping most of the stock. The pens were still full of pounded straw and muck.

"That'll make good banking material," he decided. The hired man looked dubious.

"Get the stoneboat and the team and we'll dig it out before it freezes hard. Needs cleaning out, anyway."

Several hours later the half-frozen muck was snuggly packed around the base of the house. It was very cold, and only a faint smell of pig manure drifted inside.

"Sam, you should have used the plain straw. That pig stuff smells awful!" Kate was nauseated at the slighest stench but had been too busy to note what the men were doing until it was too late to stop them.

"It will be frozen hard as a brick by morning, Missus," her husband assured her. "Next spring it will have sunk into the ground and you'll have some good flower beds there."

The next day the weather changed. A belated Indian summer set in and temperatures rose to the sixties and seventies. The pig manure softened and ran in rivulets about the door. The stench inside was overpowering. Each day was expected to be the last of the heat, but it continued for two weeks. Casual callers stayed only a short time and left. Everyone was sick of the unbearable smell.

One night it turned cold again. Nobody regretted the passing of the dreamy days of Indian summer, usually the best of the year.

Preparations for winter, preparing fuel and meat, began in earnest. Neighbours came one morning to help. An engine sitting on blocks was belted to a huge, circular saw on a frame and set up beside a pile of poplar poles which had been weathering since the spring, behind the house. Each pole had to be lifted onto the frame, slid into place, and pushed back and forth to cut it into stove-length chunks. Each time the saw-teeth bit into the green wood a screech, like a banshee, rose, followed by a chorus of agonized howls from all the dogs.

All morning the work went on — thud, thud, screech, howl — thud, thud, screech, howl — with nerve-shattering regularity that ricocheted on the inside walls of the house.

"When are they gonna finish? I hate woodcutting days!" The children complained and the baby howled. There was no place to escape from the sound.

"Never mind the noise. The job will be finished by noon," Kate consoled the children.

"Buster, go get me lots of wood and when you've filled the wood-box, I want you to pile a heap by the vat outside. They're killing pigs this afternoon and Dad wants lots of hot water."

"Are you gonna kill the old sow like you said you would? Are you, Mama? She's still there."

Kate looked startled at the inconvenient memory her children sometimes showed.

"Never mind that now. You get the wood piled up, and the rest of you stay here and keep out of the way of the men, now mind!"

"Keeping out of the way" warnings were Kate's method of protecting her family from the unpleasantries of butchering and animal husbandry, those grim necessities of farm life.

"I've seen lots of pigs killed," Buster chewed on a straw. "It ain't nuthin' at all."

Amy and Edith, his constant shadows, looked interested. He sized up the woodpile and the distance to the vat.

"I tell yuh what. You two help me with the wood and I'll show you where we can hide and watch — if you like."

Seeing the death of a pig wasn't as important as getting their big brother's approval. He always made things more interesting, even if it meant work.

The children started carrying the wood, block by heavy block, to the vat, while Kate got the fire going underneath. She dipped pails of water from the barrel outside the house, enough to half fill the vat.

The children waited until their mother disappeared inside. Then they slipped down to the barn where the pigs were now penned for the winter. They hid behind the ladder to the loft, amd peered through the rungs.

The hired man came into the barn with a rifle. Three pigs stood in a row grunting, snouting the trough and looking up in innocent expectancy. The man levelled the gun between one pig's eyes and pulled the trigger. A surprised look crossed the animal's witless face. Quickly an open razor was drawn across the pig's throat, slitting it from ear to ear. A bright red line, like a scarlet collar, appeared. Blood gushed from its mouth, and it fell forward into the trough with a strangled, coughing grunt. The other two pigs instantly met the same fate.

"I'm sick! I'm sick!" The girls backed up, ran out the back of the barn, and made for the house.

The dead animals were trundled by wheelbarrow to the vat, dumped into the hot water, and sloshed up and down. Several men scraped the bristles from the hides. Then the pigs were disembowelled. The stiff white carcasses were then hung by tendons in their heels on a pole high enough to be out of the reach of the dogs, and left to drain.

The neighbours who helped with the killing stayed for supper at night. The kitchen was filled with blue smoke from the frying liver. Pails of liver sat about, which the neighbours would take when they left for home.

"What's the matter with you two?" Kate peered at two listless and unusually quiet girls.

"Aren't you going to eat your supper?"

"I don't like liver." The pails of dark red meat and the memory of the stunned innocence of the pigs as they died, was too much.

"Me neither. I don't want any supper." And to their mother's surprise, they went quietly off to bed.

Within a week snow covered the ground and the cold began. Wind beat at the frosted windowpanes, and the snicking of sleet heralded the first blizzard of winter.

"Eaton's parcel! Eaton's parcel!" The children listened to the sound of the sleighbellls and the clatter of horses outside, announcing Grandpa Mark's arrival from town.

He came in dragging a huge brown parcel, banged his feet on the floor to knock off the snow and get his circulation working, and flipped off the ice that had caked around his moustache. He took off his yellow dog-fur coat and warmed himself in front of the fire.

Kate slit the brown paper and pulled out bright red sweaters, high top felt boots and sweet-smelling fur lined moccasins with beaded flowers on the toes.

"Winter's just lovely," Hazel shouted above the noise, "all these nice things to wear!" She rubbed soft, fleece-lined sleepers against her cheek. "Besides, we don't smell the pigs any more. I sure hated the smell of those razorbacks!"

Her grandfather, warmed, lit up his pipe.

"Ay, 'ee may not like the smell o' they razorbacks, but 'ee don't mind the smell o' greenbacks, do 'ee?"

The old man puffed some more, smiled, and settled back to watch the fun.

Chapter 31

Kate woke up.

Frosted windowpanes blotted out faint rays of the morning. A few practice yelps from the baby's cradle warned her of full-scale howls to come. They failed to rouse her aching limbs to their accustomed action. She turned her face to the wall.

"I've seen nothing but hired men, dogs and kids for six weeks!"

Already the household was stirring. Grandpa banged doors and rattled stove lids below. The cheery sound of the crackling fire was drowned out by sudden screams.

"Mama! Mama! Amy's burned her bum on the heater!" Edith shouted up the stairwell. She raced up to the bedroom, happy to be there first with the details.

"She backed into the stove before she got her cloe's on!"

Amy, howling, limped into the room. Bottom bare, a faint red diamond shape was fast erupting on one rosy cheek. The baby, now in full cry, joined the uproar.

"Look, Mama, look. She's gonna have T. Eaton Co. right on her backside! Just like brands on the calves!" Amy craned her neck, trying to see her burned behind.

"Edith, you take her down and put some of this on the burn." Kate reached for a tin of Zam-Buk on the dresser. Amy opened her mouth and bellowed.

"Enough of that, now. You won't die of it. Downstairs, all of you and get dressed. I can't hear myself think!"

Sam came in with tea and a bottle of warm milk.

"I feel as if I hadn't slept a wink!" Kate warmed her hands on the steaming cup and lay back on the pillows. Sam sat on the edge of the bed and put the bottle in the baby's open mouth.

"I think you need a change, Kate. Too bad we can't get over to Wapella for a few days, for a visit to the Gordons."

Kate put down the cup and sat up.

Sam leaned towards the window and listened. "The wind's getting up. Doesn't look like a blizzard yet, but you can't tell." He joggled the cradle gently with his toe.

Kate swung her feet to the floor and yanked at her stockings, the cooling tea still untasted. "We can stop at Dean's if it gets too bad. Speak to Father about the children, will you, Sam? I'll pack some things."

"Willie, get the team and cutter out for Dad." Kate glanced out of the window at the blowing snow.

"Mary, get some of the robes off the beds. We'll need them for the baby. It's more than twenty miles to Wapella." She bustled about, packing bags and giving orders.

"See you behave, now, for Grandpa. And Buster, you be sure to get those cows milked on time. I don't want them drying up until the others come in."

She turned to Grandpa. "We won't be gone more than three days, Father. I hope you don't mind keeping an eye on things."

Grandpa stuffed his pipe with tobacco, tamped it down and lit it.

"Don't 'ee worry, Kate. I be watchin' they. Change'll do 'ee good."

The children loved the freedom as much as the parents did when Grandpa was left in charge. There was no deadline for dishes. Unmade beds were just as comfortable as made ones if one were careful not to disturb the blankets too much. Grandpa didn't mind how often candy was made or how much sugar was used. He willingly cooked anything, anytime.

"Grandpa! Grandpa! Make pasties. Make pasties!"

Unwashed dinner dishes were put to one side. Grandpa, between puffs at his pipe, rolled out the dough on the food-spattered oilcloth.

An hour later nobody recognized stray bits of tobacco in the steaming Cornish treats. Nor was any notice taken of spilt flour tracked from table to stove.

Privacy was unknown in the crowded household. Only Grandpa's locked closet remained a mystery.

"Grandpa! Grandpa! Show us your pipes. We wanna see your pipe collection."

Resistance weakened and the old man unlocked the door. He was wedged in by pressing bodies.

"Lemme in! I got here first!"

"You're on my foot! Grandpa, make him stop shoving!"

"Gor dairney all! Back! Back, I say!" Grandpa shouted and waved his hands above his head in pretended exasperation.

Lovely smells wafted out. There were odours of good cigars, the new smell of his mackintosh, and a faint whiff of lavender from their dead grandmother's unopened basket trunk. There were other emanations, the source of which could only be imagined. These were things acquired on ocean voyages or during his other life long ago.

He brought out a large flat box. A collection of pipes, each lying in a fitted groove on blue velvet lining, filled the box.

"Grandpa! Grandpa! Let us smoke your pipes. Learn us to smoke, Grandpa. Learn us to smoke!" Buster, always pushing to be like the men, was more anxious for the experiment than the others.

The old man selected several pipes, a Meerchaum and some with straight or curved stems. He carefully filled each with tobacco.

"It'll make 'ee sick, now. I'm tellin' 'ee, 'ee'll be sorry. 'ee won't like it."

With a pretended show of resistance, a pipe was relinquished to each child.

"Don't 'ee swallow, now. Just puff."

Buster listened and followed instructions. His pipe curved down well below his chin. He grasped the bowl in his hand, clenched the stem in his teeth, and puffed. He was rewarded with a satisfactory and continuing cloud of smoke.

"I can do it, Grandpa! I know how!" he shouted between puffs. He suffered no ill effects.

For the other children, varying degrees of success or failure followed. Some puffed, some blew, and all swallowed smoke. Except for Buster, everyone began to turn a pale saffron colour. Grandpa sat back and puffed contentendly at his own corncob. He relit pipes and watched proceedings with no air of censure. He was a child again, happy to be part of the conspiracy.

The next morning the noisy uproar caused by school preparations was traumatizing. Grandpa's presence didn't help.

"Gor dairney all! Do 'ee 'av to sleep on the job? 'Urry up now, else 'ee'll all be late!"

He stood dead centre of the kitchen waving his arms above the heads of the frantic school-children. Then he pulled on a worn yellow dogfur coat, buttoned the only two remaining frogs at the top, and tied a child's scarf around his middle.

"Do 'ee look sharp, now! I'll 'av the team up dreckly!" He marched out, scattering dogs, cats and children before him, and leaving the kitchen door wide open to the freezing blasts.

"Where's my mitts?"

"Who took my toque?"

"My sweater's soaking wet!"

Ruth tackled one uncombed head after another.

"Why don't you put your things where you can find them, Buster. You should have hung your sweater up last night. Hurry up, now. Grandpa gets mad when he has to wait."

Hazel buttered slices of bread and jammed them into the lunch pails. Mary put bricks in the oven to heat and threw more wood on the fire.

"There's the team! Hurry!" The children rushed out, books and lard pails clutched in mittened hands. A black furred buffalo robe and several cowhide ones were thrown into the sleighbox. The bricks from the oven were wrapped in newspaper and put in for foot warmers.

"Whoa, Bess, sooh, boy!" Grandpa, corncob pipe unlit in his mouth, stood up front and soothed the restive team. Everyone piled in and they pulled the blankets over their heads.

"Look sharp, will 'ee. Look sharp, I say!" Like a ringmaster directing a show, he waved his willow switch aloft and joggled the frozen lines. The team leaned into the harness and the rig lurched onto the trail. Indifferent to his repeated threats, they trotted over clear icy stretches and slowed to a floundering walk through fresh drifts.

The muffled sound of singing rose from the blankets.

"The maple leaf, our emblem dear
The maple leaf forever!
God save our king and. . . ."

Grandpa smiled and urged the horses on.

"The maple leaf . . ." The song faded.

A mile along the road the wind had whipped out the last bit of heat from the foot bricks. Cold penetrated every boot-toe and mitten-tip. Tempers sharpened and sniffles began as the rig neared town.

"I'm cold."

"Don't move. You're letting in the wind."

"I wish Grandpa would drive us every morning. Then we'd be early."

"Yeah! It was you girls. I always got the team out on time and had to wait for you!"

Willie, who always drove in good weather, resented his demotion to the bottom of the sleighbox with the little kids and the girls.

They reached the school door just as the teacher came out to ring the brass hand bell.

"After the opening prayer, you Marks children may sit around the stove and get warm," she said. "You have a long way to come." The children sat, faces scarlet from the roaring furnace. They pushed their boots under the tin shield which surrounded the stove and acrid smell of scorching leather mingled with the odour of drying wool. Chillblains, brought on by the sudden changes in temperatures, made their feet ache and were almost as painful as the cold.

In the afternoon, temperatures outside rose. The rutted trail became soft and slippery. Nobody needed the robes. Everyone stood up and wanted to drive.

"I know how, Grandpa. You don't need to hang onto the lines." Hazel clucked at the horses like the hired men.

The irregular yanks on the reins as the children took their turns alarmed the usually docile team. As they topped the last rise before home, one horse reared sideways, staggered over a snowbank and floundered to a stop. The other horse followed, plunging and rearing into the deep drifts. The entangled harness further alarmed the team. They plunged and snorted, with rolling eyes and grunts of panic. At the first lurch the lines slid out of small hands and over the side of the box. Grandpa swung around, tried to grab the lines, and lost his balance. The sleighbox tipped, twisted and fell on its side, slewing books, children, lunch pails and hay over the road. Grandpa catapulted into the air like a rubber ball, rolled over and over and landed in a snowbank. His silver spectacles hung from one ear. His pipe was broken and a steady stream of Cornish curses issued from his throat.

"Grandpa! Grandpa! Are you alright?" The children crowded around him, alarmed at his ashen colour and the peculiar words he was saying. Willie grabbed for the horses' halters, pulled firmly and talked soothingly to quiet them.

The old man stopped his noise for a moment and looked completely puzzled. Then he struggled to his feet, waved his arms in the air, and shouted with his customary vigor.

"Gor dairney all. Gor dairney all!" and normalcy was again established.

During the evening the wind rose in gusts. It moaned through the creaking telephone wires attached to the house, and echoed eerily in Grandpa's tiny room. He sat on the edge of the bed, feet scarcely touching the floor. Wisps of smoke curled around his head. He puffed, listened, and smiled each time a fresh blast struck the house.

Nailheads by the window were furred with frost. Silvery ferns and leaves snaked up the rattling panes as the cold outside increased. Hazel, Buster and Edith pattered into the room in fleece-lined sleepers. Amy tagged along behind, sucking her thumb and twisting a hank of her hair.

"Read us a story, Grandpa. Read to us, will ya?" They climbed up on his bed and snuggled down deep in his blankets. Their tousled heads shone in the tiny circle of yellow light fom the coal oil lamp.

Amy took her thumb out of her mouth and shoved a well-worn copy of Les Miserables towards him, one of two books he kept for the children.

"Read us about Cosette, Grandpa — where she didn't have no clothes or nobody to make her porridge."

"Never mind her, Grandpa. You read us about her before. Read about Daniel. I like Daniel." Buster pushed a book of Bible stories towards him. It was opened at a page which showed Daniel strolling serenely, with his hands on the heads of two adoring lions.

"Grandpa! Grandpa, it's my turn to choose. I want Solomon and the babies, Grandpa." Hazel hung onto one page of the old book. It showed a soldier holding a baby aloft by one leg, obviously about to chop it in half with a big sword.

"Do 'ee 'av a little patience, will 'ee! I'll read 'ee all some'at 'directly."

He rescued both books from clutching hands, adjusted his silver-rimmed glasses, and opened the bible book at the story of Moses.

"Don't read about Moses, Grandpa. I don't like Moses unless he's in the bull's rushes," Edith complained.

"Don't 'ee like Moses, now? What for?"

The old man's voice always had a special ring when he read of the great leader's struggle to lead the stubborn Israelites to the promised land.

"He was always saying he got those spooky messages from God, an' I don't think God liked him neither, or He'd of let him get to the promised land, too. Read about Jesus. I like Jesus stories."

Grandpa read the story of Christ at the wedding in Cana. Gradually each child began to yawn and snuggled closer. Grandpa closed the book.

"Look sharp now, will 'ee. Go to bed now."

Household noises died. The lamp wick guttered and went out. Clouds scudding across the moon outside made moving shadows across the wall. The wind hammered on, now gently, now in gusts.

Sleighbells tinkled and the dogs began to bark.

"They be home again." Grandpa smiled, and slept.

The next morning he got up late. Sounds of frantic activity rose from below as Kate waded into deferred tasks. Edith banged at the door.

"Grandpa! Grandpa! Lemme in. I want to ask you a question."

The old man opened the door.

"How come Jesus made wine for the wedding, Grandpa? Wine's booze, isn't it?"

"Whudd'ee say, girlie? Whudd'ee say?" He listened with more attention than usual.

"About Jesus giving booze to the people for the wedding. Mama says booze is the downfall of man. Wine IS booze, isn't it?"

Grandpa's eyes twinkled. He lit his pipe with great concentration.

"Go ask your mother," he said at last. " 'Er be tellin' 'ee better nor I."

He trotted downstairs after the child and dodged behind the kitchen door, straining to hear.

"Mama! Mama, why did Jesus make booze for the wedding?" Edith tagged after her mother, pulling at her moving skirts for attention. "If wine is booze and Jesus gave booze to the people why is it bad? Jesus must have thought it was all right. Wine IS booze, isn't it?"

Kate stopped and stared at the child.

"It's — we — it's . . ." she began.

Grandpa's face, guileless as a baby's, appeared around the corner of the door.

Kate chose the quickest way out.

"Never mind the wedding wine! I'll wine you if you don't get at those dishes!" She shoved a dishcloth into the child's hand and bustled into the dining room, with only a backward glance at Grandpa's innocent face.

"Drat that child. Where DOES she get her ideas?"

The old man trotted upstairs again, chuckling to himself. He came down again later, with a special shine in his eyes and smelling faintly of cloves.

CHAPTER 32

Spring came and with it, change. Edith and Amy sensed it most in Buster. He treated them with greater condescension, left them the filling of the woodbox, and spent more time in the barn helping with the "chores."

They yearned to assume his possessions, step into his shoes, and enjoy his privileges, a family promise that was constantly violated. His shoes were boy's shoes which they couldn't wear, his privileges widened because his preferred habitat was the barn, and his acquired possessions, its produce thereof. He now owned a pig which he could claim money for at shipping time.

He took to spitting in an arc like it was tobacco juice, hanging his clothes on the high nails like the grownups, and smelling proudly of manure like the hired men.

He pushed on the door of adulthood in other areas, too.

One June morning in the midst of the "we-don't-have-to-go-to-school-it's-Saturday" confusion, he rushed into the house, washed his face and hands without being told, and donned his best Sunday suit, without being noticed. Edith and Amy followed him, fascinated.

"I own a horse now. Nobody rides him unless they ask me first."

Old Jim, on whose patient back they all learned to ride, was tied to a post by the kitchen door. His black pelt shone like slipper satin, from gentle ears to ratty tail, and his hocks had a dampish look. They'd been washed and scoured clean of stable manure at the well.

"I gotta feed him and give him his oats — not too much because it makes him too frisky." Buster waited to let these snippets of wisdom be absorbed.

"I gotta currycomb him, too." He gave his own hair a swipe with a wet brush. "And clean out his stall." He re-buckled his knee-pants and buffed the toes of his boots with the sleeve of his coat. Then he marched out of the door, stopped and stood back to examine his new property for flaws. His sisters followed close behind, and looked for flaws too. There were none.

"He's clean as a whistle!" and Buster undid the tie rope and climbed on old Jim's back.

"Gosh, I nearly forgot something." He slid off and tossed the neck reins to Amy. Edith followed him into the house and up the stairs to the door of Grandpa's room. He went in and eyed Grandpa's collection of hats carefully.

"Grandpa, kin I borrow yer Christie Stiff?" I'm going up to see Vera Marshall," he shouted, with the dedicated purpose of a seasoned Romeo. "They're the new people just moved onto Patterson's place, and she's awful pretty."

At that exact moment Kate came out of the other bedroom.

"Go and take off that suit this minute, young man," she scolded. She moved into Grandpa's room. "You don't put on your good suit until tomorrow. Get into your overalls and put the harness on old Jim. I want you to bring up a barrel of water on the stoneboat." She pushed her son out of the room and shut the door. The sound of laughing followed them all the way down the stairs.

"I wonder what's so funny?" Edith remarked to the grounded suitor. "They sure laugh at some funny things."

That night visitors came. The children were sent to bed early. Hours of exciting adult conversation, peppered with laughter, floated up through the grate, a lacy iron square set in the bedroom floor to let up heat in the winter. Edith crept out of bed, lay down on the floor beside the grate, and listened to the grownups' talk.

"They're robbing the farmers right and left" meant the talk was concerned with railroad prices and financiers in the east.

"We'll put the scoundrels out next election" meant the conversation was political. (The incumbents were always at fault, no matter which party held power.)

Conversation was switched to Mrs. MacDonald, lately widowed, who had been willed something. It was the last bit of talk Edith heard as she drifted off to sleep with her behind in the air and her ear pressed to the cold iron grate.

When Grandfather Marks came to live with the family he brought his fat sorrel mare, Darling, with him. She shared a stall with old Jim, the children's black pet horse. She was as much loved as Jim, although she was liable to lay back her ears and make teethy threats, or switch her long orangey tail impatiently if she was urged to trot when she wanted to walk.

When things got too frantic at Sam's, Grandpa locked his clothes closet, hitched up Darling to her smart new buggy and drove off to visit his son Will.

"I be off to Will's today," announced the old man the next day.

"I be taking Edith with me."

Darling trotted briskly out of the gate and took the winding Indian trail, deep-cut by generations of nomads who had travelled over the plains from the Moose Mountains to the Pipestone Valley. The horse moved at an even clip down the grooved centre of the trail. The sun glistened on her silky pale-sorrel hide, her smooth rounded rump swung rhythmically and her heavy hooves kicked up whorls of dust as she jogged along.

Suddenly her ears pointed forward. She came to a stiff-legged halt, then bolted sideways. The old man pulled firmly on the lines and spoke soothingly to quiet her.

Around a little knoll beside an outcropping of brush appeared a bell-shaped framework of plaited willows. It was the exact size of a little girl's playhouse.

"We bain't touch it," warned the old man, "it be an Indian shelter. They be usin' it when they travel by 'eer." Edith watched with regret until it was out of sight.

"Grandpa! Grandpa! Can I drive now, Grandpa? Can I?" Only when Buster wasn't there was this privilege available.

" 'Ee be careful, now," Grandpa handed over the reins.

The warning was wasted, for the horse dragged to a walk.

Meadowlarks sang cheerily from fence posts, blackbirds trilled from slough reeds and mourning doves cooed plaintively as they passed. A weasel, glowing winey-brown in a new spring coat, slipped by in front, scarcely moving the grass, his ferocious beady eyes glistened and intent on some mystery errand of his own. Perfume of bursting buds, fresh grass and Balm of Gilead mingled with the smell of dust. Edith held the reins still and let the horse walk and was content. The old man's eyelids drooped and he nodded, waking with a jerk and immediately nodding again.

"What's 'willed it' mean, Grandpa?" Edith shouted to wake him up, and to be heard above the creek of the wheels. Darling gave a jerk of surprise at the sudden noise and began to trot. Grandpa woke up with a start, pipe still clenched in his teeth.

"Whudd'ee say, girlie' Whudd'ee say?"

"It's about the Widow MacDonald," she explained. "They said she got a 'willed-it' last night. I thought Mr. MacDonald died. Did she get a 'willed-it' to take his place, Grandpa?"

The old man unclenched his pipe to laugh. He laughed a lot lately at things that didn't seem funny. Then he explained the meaning of wills and inheritances.

" 'Ur be getting' 'ur 'usbin's varm."

"All the cows and pigs and the horses too, Grandpa?"

"Yes, all uvit."

Edith flapped the reins on Darling's back and made a kissing sound like Buster did when he was driving. An idea had just struck her like a clout on the head.

"Grandpa, who will Darling and the buggy belong to when you die?" she yelped over the rattle and the creaking of the wheels.

"Whudd'ee say, girlie? Whudd'ee say?" Her grandfather looked startled.

"Who will get Darling when you die? Somebody's got to get her. Will her to me, Grandpa? Will her to me, please?"

He took his pipe from his mouth, turned to stare at his grand-daughter for a moment, and then laughed uproariously without answering.

"She's got to belong to someone," pleaded the little girl. "Buster's got Old Jim. Will her and the buggy to me, Grandpa. Will you?"

"Yes, 'ee may 'av 'er," he promised at last. He kept laughing and muttering to himself for some time.

The spinning wheels slipped over the miles and they entered a cool avenue of greening trees, their lacy branches arched together overhead. Pincherry fruit, just beginning to form, rained tattered white blossoms like perfumed snow as they passed. Darling's gait quickened, her ears poked forward and her flying hooves beat a happy tattoo on the sun-spattered black loam beneath the trees. They were almost there.

The visit was a short one, for Edith clamoured continually to go home.

"Hurry up! Hurry up!" she shouted at the horse on the home journey. She could hardly bear waiting for the time when she could brag to her brother about her new fortune.

They weren't back home fifteen minutes before all the children knew.

"Wills are what people do with their stuff when they die, and that's why I'm to have Darling and the buggy. You can ask Grandpa, if you don't believe me."

Buster looked like the bird who arrived too late for the worm. He could see Old Jim and Darling as a nice team under one ownership — his. He wasn't entirely convinced until Grandpa verified it in a vague manner. Edith queened it over everyone with as much joy as a sweepstake winner.

The children ate supper of bread and milk; then rushed up to Grandpa's room.

"It's true, Grandpa, isn't it? You willed me Darling when you die!"

Six children crowded round him, five of them looking gypped. Grandpa sat on the edge of the bed, puffed at his pipe, and laughed.

Willie grabbed at a big silver watch which hung on a heavy chain around Grandpa's middle. A sailing ship was etched on the back of the case and "Plymouth 1880" was written below it.

"Will me your watch when you die, Grandpa, will you?" Willie shouted, "will me your watch!"

"Yes, 'ee may 'av it!" Grandpa chuckled. The late-comers flew about the room grabbing everything moveable that might be of future value.

"Will me this, Grandpa, I want this!" Buster grabbed a leather collar case full of stiff starched collars. Hazel lifted a small framed picture of the lighthouse on Plymouth Hoe off the wall, and claimed it. Amy clung to a shaving brush in a porcelain dish half full of caked soap.

Willie took the watch and slipped the chain out of Grandpa's pocket. He tried it in his own overall pocket for size and weight, then pressed it to his ear for sound effects. Then he leaned over the old man's shoulder and gave him a close scrutiny.

"How have you been feelin' lately, Grandpa?" he asked seriously.

The same thought struck all at once. Grandpa looked pretty healthy. The subject of concern threw back his head and roared.

"You children come right down and stop bothering Grandpa!" Kate shouted from below.

"I don't know what's wrong with him," Edith confided to her mother later, "he always laughs at the wrong things. Why did he laugh when we asked him when he thought he was going to die and his will and all that?"

"You WHAT?" demanded Kate.

Edith explained about Grandpa laughing at the wills, and Kate laughed too.

"All I can say is, your grandfather has a wonderful sense of humour. He needs it with you children."

CHAPTER 33

The winter of 1916-17 was a bitter one for western Canada. The stunning monotony of cold below-zero temperatures, coupled with the number of local boys killed at the "front", was depressingly sad. Although the calendar promised spring, March opened like a lion and continued to roar until the evening of the sixteenth. Roads drifted over. Getting to town was too difficult for the children. For a week, school had been out of the question.

A raging blizzard battered the house all day. The children stayed close to the stoves, helped bring in firewood to stoke them, and quarreled continually among themselves. At suppertime Kate became ill and disappeared into the parlour-bedroom. The children made the meal. After supper Sam held a murmured consultation behind the parlour door with his father. Then they proceeded on a mysterios errand of their own. Foot-warmer bricks were heated. They donned fur coats and wound scarves about their heads and faces. A team was hitched to the cutter and they floundered through the drifts towards town.

At one o'clock in the morning, unusual sounds of activity woke the sleeping house — the men had returned with Mrs. Lloyd, the mid-wife, in tow. She stepped from the cutter covered with snow, entered the house and removed her outer clothing. Disregarding the cold, she donned a stiff white apron, pinned on a nurses' cap, and took command of the household with tight-lipped efficiency.

"Mrs. Lloyd! So glad you've come," Kate murmured from where she lay in bed in the parlour.

"Another child! Isn't this the NINTH?" Mrs. Lloyd sniffed and scolded as she bustled about the untidy room. She dug into boxes and baskets for clean sheets and towels.

The men turned the team over to the hired man and stamped into the house. They threw off mitts and coats, and stuffed more wood into the fire-boxes.

"I'll need hot water and more heat immediately," directed the nurse.

"Mmmmnph!" snorted Sam. His black eyes snapped. He thumped down the kettle, filled it with water, and turned his back on the nurse. Her caustic comments continued, about the inadequacies of the house, the number of children it contained and the low regard in which she held all men.

Grandpa Marks, suddenly deafer than usual, brought in more wood, lit up his corncob pipe, and smiling contentedly sat down by the fire to wait.

Mrs. Lloyd sailed in and out, searching and finding every deficiency. The adequate supplies of heat and hot water seemed to disappoint her. She looked at once for other reasons to censure and found one — the old man's pipe. She scuttled over and stood over him belligerently.

"I LOATHE tobacco! It's extremely bad for a confinement case. It may have a dilatory effect on the child. I demand you stop smoking at once!"

"Whudd'ee say, Mrs? Whudd'ee say? I'll make 'ee some tea — o' course . . ."

Sam was snoring in his chair by the time Grandpa Marks came in from the kitchen, smoke-belching pipe still clenched in his teeth. He carried a tray with a pot of bubbling hot tea, a good china cup and saucer, and a plate of warmed Cornish pasties.

Mrs. Lloyd stopped from her ceaseless activities and surveyed the proffered lunch with pleased surprise.

"Why, Mr. Marks! Pasties? How nice. Did you make these your-self?"

The old man's hearing miraculously improved.

"Made 'un for the children for supper." He smiled engagingly at the nurse, puffing the while. "I'll just fetch another cup for Kate and myself."

Sam snored on, only ceasing his noise when his head lolled too much to the starboard, which woke him sufficiently to get in a more comfortable position.

By the time a bright March morning broke across the plains and a wintry sun, flanked by rainbow-hued sun-dogs, flooded the drifted snow, Mrs. Lloyd's demeanor, usually sharp and starchy as her nurses' skirts, had quite changed towards Grandpa Marks. They had arrived at a great and touching mutual sympathy.

Before the day was minutes old, Patrick Marks arrived. The influence of Irishman's day may have had something to do with it, for from the moment of birth, his fat and wrinkled little face seemed wreathed in puckish smiles. Instead of the usual lusty Marks' howls at birth, Patrick gave out what was strangely like a gurgling laugh.

The next few days were busy ones, and everything was pushed back to accommodate the new arrival. Mrs. Lloyd still maintained her manner of contempt towards the father of the new child, but she had warmed considerably towards the grandfather, and was pointedly telling him her Regina address, for she too was a "sad and lonely widow".

Another year went by and in the month of May, Mrs. Lloyd's ministrations were called upon again. Margaret Isobel, Kate's tenth child, was about to make her entrance into the world, but Grandpa Marks was strangely absent when the time arrived.

"I've got Mrs. Lloyd coming in on the train tonight, Father," Sam said. "Will you drive in and pick her up at the station. The increase should be any time now and I don't like to leave the Missus alone."

The old man, hoeing in the garden in preparation for the spring planting, looked alarmed. He took his pipe from his mouth, sputtered for a moment or two in some agitation.

"I be goin' to Will's tonight," he objected, " 'ee be 'avin' some trouble with a cow and I needs be there to 'elp."

He left hastily for Will's, leaving the garden half-planted.

Mrs. Lloyd frequently inquired in an offhanded manner as to the whereabouts of that "dear Mr. Marks".

That fall half of the garden froze from late maturing, because Grandpa didn't get back to his planting until late in June, after Mrs. Lloyd had gone.

CHAPTER 34

"Who be comin' to Will's today, Kate? I be goin' dreckly after brekfus'." Grandpa Marks came downstairs and joined the noisy group of children around the breakfast table.

"It's my turn to go. I wanna go this time!"

"You went last time. I want my turn."

"I never went once this year. I wanna go."

"Just you stop that arguing! It's Hazel's turn," pronounced their mother.

The children fought for the privilege of going to their Uncle Will's because of the different things he had to eat. At home no one was coaxed to eat. Food was plain but adequate. It was non-existent if one dawdled at noon or at night.

"Too bad. Let that be a lesson to you. You'll know enough in future to be here on time!" was Kate's answer to complaints about empty bowls and platters.

One meal only had a surplus. It was breakfast. Then there was always lots left over to share with the cats and dogs, who waited eagerly around and under the table.

Kate rushed in with a two-gallon pot of grey oat porridge, which she ladled onto each plate.

"Just you watch how much sugar you put on it, now."

"I want cream on mine. Uncle Will always lets me have cream."

"I don't like husks. This porridge has husks."

"I don't like porridge. Can't I have toast. The McClements always have toast made over the coals in the fire. They have store-bought marmalade, too."

"Never you mind McClements. You eat your porridge or nobody will go to Uncle Will's."

Kate peered down the table to find out who was complaining.

"And leave that cream alone or I won't have enough for my churning."

Everyone took an extra splash and pretended not to hear.

"Can't I have more sugar on mine?"

"No. One spoon's enough." Her objections were weakening. "I learned to eat porridge without sugar."

"Yes, but Mama, you had maple syrup in Ontario, and that's lots better than sugar anyway."

"Eat your porridge," and Kate passed the sugar bowl around again.

Hazel's trip to her uncle's that day was talked about and regretted by the other children for months.

After she left with her grandfather, Edith and Amy settled down to their usual occupation. Playhouses were built with a few sticks, clothed with imagination. All day they played endless games of "Rich", a pre-schoolers one-up-manship which was concerned mostly with food and required quick and fertile imaginations.

Kate's old hats were dragged out and decorated with hen's feathers. They put on the dead grandma's black taffeta dresses, stuffed them up front with old socks and towels, and tied them around the middle with belts, which gave them the appearance of pouter pigeons, and were quite satisfactorily grown-up looking. Dandling their dolls over imagined cups of tea, they bragged about the supposed affluence of pretend husbands.

By evening their refinements of wealth had reached satisfactory heights.

"Eustace brought me two hired girls yesterday, one to make sugar cookies and the other to make chocolate cake." Edith burped her doll complacently and waited.

Amy rocked a bit and said, "We went to Langbank picnic last week and I ate ten ice cream cones in a row. Jim always has lots of nickels when we go to a picnic."

"Well, Eustace always lets me have real cream on my porridge. It's good for my complexion, he says."

The sound of buggy wheels in the yard announced Grandpa's return from Will's with Hazel. It interrupted the game, and the two children rushed outside.

"What did you get to eat? What did you get to eat?" they shouted. Anything not edible didn't count.

"Come and tell us what you had!"

"Well, I had an orange and gumdrops and pasties, but the best of all was cornflakes! You can't imagine how delicious they are — better than anything you ever ate in your whole life!" Hazel climbed down from the buggy. Her sisters listened bug-eyed to a minute, mouth-watering description of yellow-orangey crinkly bits of the food of heaven in never-empty bowls, a-swill with cream, brown sugar, and a slice or two of banana.

"I ate cornflakes all the time, and Grandpa let me eat all I wanted."

"Are they as good as Mama's gingersnaps?" Kate frequently made inspired panfuls; their delicious molassesy perfume could be smelled a mile away. At the first whiff of it the children raced towards home, stomachs rattling in joyous anticipation.

"A hundred times gooder!"

"How about peas? Or liver?" Amy had peculiar preferences in food.

"A million times better!"

"What about chocolate bars. Or crackly nut popcorn? Or ice cream? Nothing's better than ice cream."

"They're ever so much better!"

The children begged for descriptive replays until Hazel wearied of the subject and wandered off. The other two returned to their playhouse, tea and one-up-manship.

Edith poured another cup of invisible tea, her mind on the quickest way to get back to her uncle's. Amy delivered the coup de grace and won the contest hands down.

"We have cornflakes every day," she asserted, not waiting for her sister's turn to brag. "This morning the box got knocked on the floor and the whole thing spilt. I didn't even make the hired girl pick them up. I just swept the cornflakes right out the door!"

Everyone was now frantic to go to visit their uncle.

"Grandpa! Go to Uncle's again, please. Go tomorrow, will you?" The children plagued the old man so assiduously that he gave in at last. Three days later he hitched up his horse, and with Edith and Amy, headed north. They were almost content to let him drive all the way. Darling kept a steady trotting pace for him.

When they arrived the children leaped from the buggy and raced into the house.

"The cornflakes! Where are the cornflakes?" they shouted to Uncle, who was just reviving from an afternoon nap.

A totally empty box sat on the shelf, and no trip to town for groceries was planned for two weeks.

They both burst into tears. It was a crushing blow, and only partly alleviated by two chocolate bars and a bag of store-bought cookies.

CHAPTER 35

"Gee-Hup! Whoeehhh!" Mike Hayes pulled a road-weary team, lathered with sweat, to a halt outside the kitchen door.

"Missus," he called from the front seat of the democrat.

Kate, followed by a contingent of young ones, slammed open the door.

"Sam and the old man will be along soon." He grasped the lines in one hand and fished in his pockets for chocolate bars.

Shading her eyes from the glare of a hot August sun, Kate peered up at Mike anxiously.

"Did they buy the Ford, Mike?"

"Whooehhh. Steady now!" Mike soothed the team. "Here yuh are, kids." He tossed out four bars.

"You get on the barn roof and you can see them coming. They've got a

surprise for you. Hurry!" He guided the horses towards the barn without answering Kate's query.

For months, cars had been discussed endlessly among the family, with visitors, or with neighbours. The war brought prosperity to the prairie wheatlands. Every bushel was needed, and prices shot up. Top prices for pigs and beef cattle made financial success seem imminent.

"These steam cars sound good — cheaper to run." Sam examined an advertisement, "I don't think they'll last, though — too heavy." He gloated over the pictures and read the descriptions aloud to Kate, as if she was enthused too.

"I never seed they ones much in Lunnon," remarked Grandpa Marks, the world traveller. "They Fords are the thing — seed lots of' they. They be easy on fuel — and cheap!"

"Nothing's cheap if you don't need it and can't afford it," snapped Kate. She rushed from the kitchen to the pantry, raising objections as she went. "You don't know what prices will be next year, or even if there'll be a crop."

"Cars be the comin' thing. Kate.'E'll like to ride in un," Grandpa assured her.

"Mmmph! We're too fond of pleasure now — it's not good for the children. It'll be on the road all the time —"

"'Ee can be in town in fifteen minutes, stead o' an hour," the old man reasoned. He looked up anxiously at his daughter-in-law. She picked up dishes, slammed them down and rattled pots and pans angrily.

"Look at this fancy one," said Sam enthusiastically. "Ever see one of these overseas, Father? And Kate, just listen to this." He raised his voice above the crashing of the crockery, and read out detailed descriptions of velvet upholstered tonneaus and interiors with silver fittings.

"You could ride up front, Kate, and we'd close the glass between — lots of room for the kids in back in that one, eh, Father?" and they roared together over the long, black, chauffeur-driven car.

"Better look at these Fords, though. Best kind for this country." The two men bent over the pictures again, and murmured happily between themselves.

"Best place for that money is in the bank — we can't afford —" Sam read out bits from time to time as if Kate were in complete accord to the outlay of money.

One hot morning a few days later Sam announced, "We're going to Wapella —" and muttered something about repairs for the binder. He set off with Mike and Grandpa in the democrat. They all looked quietly pleased. Kate knew she had lost the argument, as expected.

"Somebody's got to put on the brakes or those two would buy everything in sight." She frowned, not bothering to wave as they went out the gate.

The dust from the wheels had barely settled when she stopped her work, wiped her hands on her apron, and with a smile of pleased satisfaction, cranked the phone.

"Yes, the men have gone to Wapella to look at a Ford car," her voice bubbled pleasantly to Mrs. Potter, a telephone friend. "Such an improvement over the democrat — the children get so dusty on the way to church — and fast, too. We live in a wonderful age."

The next day Mike came home alone and made his announcement. Hazel, Buster and Edith, with Amy toddling behind, raced to the barnyard, almost too excited to eat their chocolate bars. They climbed up a strawstack, jumped to the roof of the cowshed which leaned against the barn, and then up the slanting shake roof to the top of the barn, where they perched like birds along the ridgepole. The brassy August sun shrivelled and dried the rivers of sweaty dust on their faces. They checked bare feet for slivers from the climb over the shingles, ate chocolate bars, and trained their eyes to the north-east across the softly undulating miles of treeless, copper-coloured fields glittering in the late summer sun.

"Everything's turned to gold like King Midas in 'The Golden Touch'."

"Aw, g'wan," Buster, the realist, scoffed. "That's just the forest fires out in B.C. that makes it look like that. Grandpa told me."

"Everything LOOKS like it's turned to gold — like we're rich." Amy stuck to opinions regardless of reason.

"Rich people all have cars now, and if Dad brings home a car, then we'll be rich!"

"If we're rich then we can have all the ice cream cones we want." The children considered anything an advantage only if it was edible.

"Waa—all, most likely we won't be thinking about ice cream when we're riding in our car." Buster pondered the possibility and squinted across the miles. "And I'm gonna drive. I'm gonna drive twenty-five miles an hour." He recklessly steered an imaginary vehicle.

"Nobody can go that fast," Hazel corrected him, "besides, you don't even know how to start it."

"Well, I guess I do so," and he gave a complicated explanation involving sparks and cranks. He was well versed in the vagaries of car culture, having spent half a day in Kennedy at the Markle's, who owned one.

"They'd prob'ly be here now, but they've likely got a flat tire." He looked superior and everyone was too hot to argue.

A tuft of dust, like a thin curl of smoke, materialized on the horizon.

"It's just a little whirlwind."

"It's just dust."

"No! Hey! It's them! It's a car. They're coming!" The wish turned into a cloud which moved along the horizon at an unbelievable speed, and ballooned in volume as it approached. A faint, sullen, mechanical beat increased into the unmistakeable engine thump.

They screamed, slid and jumped down over shingles and stacks, with no thought of slivers in bottoms or feet. Across the barnyard they raced, under the barbed wire fence, through the waist-high thistles and nettles, and across a stony rutted gully towards the Indian trail east.

At the top of the next rise they stopped. With a snort and a roar and a backfire or two, the shiny, black, trembling monster, canvas top snapping and vibrating in the wind, drew up and stopped beside them. Father and Grandfather, pleased as schoolboys with their first bike, sat stiffly upright and smiling in the front seat. Sam clutched the steering wheel with the stiff-armed grip he used on a fractious team. Grandpa Marks hung onto his homburg hat with one hand and with white-knuckled intensity, clung to the side of the seat with the other.

"Get in the back, kids. Hurry!" They jumped on the running-board, opened the gate-like door, and piled in — that is, all but Buster. He was down examining the back wheels.

"Hey, Dad! This ain't a Ford, is it? That's not the name on the wheel here!" He had just started school and was showing off his ability to read. "This word starts with an M!"

"It's a McLaughlin. Now jump in, or you'll walk!" Sam made some adjustments with his feet, and this miracle of the industrial world shuddered to a start and moved slowly down the road.

The car gathered speed, hitting and bumping over stones in the gouged out ruts of the trail. Sam wheeled it out onto the grass by the trail side. The rest of the way home was a cloud ride for the starry-eyed children with their tousled heads of hair streaking out in the wind. They clung to the bar along the back of the front seat and bounced along on the buttoned down leather cushions.

A flock of heat-stunned fowl pecked disconsolately in the yard, or stretched and dusted themselves in the shade of the hen house. The car suddenly appeared at the gate and rumbled towards them. With a burst of panic the hens flew into the air, scattering feathers and screeching mindlessly. Ducks waddled frantically into the path of the car, and the geese, with shreds of their usual dignity left, honked wildly and tried to escape. From the cool side of the house, the dogs heard the noise and leapt into action. They ran and snapped at the rubber tires, while the children screamed joyously at them from the back seat. The car trundled up the hill and stopped in front of the kitchen door.

"Tell your Mother to come out." Sam still gripped the wheel, and his father, bolt upright beside him, sat waiting.

"Mama! Mama! The car's here. It's a lovely big McLaughlin!"

Kate, prepared to grudgingly accept a Ford, stopped in the doorway. The car suddenly seemed twice as expensive looking and imposing as it had down the road. The children jumped out and rushed to her, pulling at her skirts towards the car.

"I thought — a McLAUGHLIN! — You said — I thought you were going to buy a Ford?"

"Come on for a ride, Kate!" The two men began to look guilty.

"It's a custom built job. Rides like on a cushion — last twice as long as a Ford." Sam roared the engine.

"C'mon to town, Kate," he coaxed.

"How much did it cost?" Kate was determined not to be put off.

"Father's paid for half of it, and it was only $1,400.00."

"ONLY fourt —"

"C'mon, Kate, we'll go to Langbank and back!"

"I've got to get supper!!" She marched into the house, back straight and head high.

Some days later Kate answered a knock at the door.

"Is there something I could do?" she queried. "Sam's out in the field right now." It was a carpenter from Wapella.

"I'll just unload my stuff here. I came to build the car shed for Sam." He dumped off a suitcase and drove off to the barn, the tools of his trade; kegs of nails, hammers and saws protruding from the back of the buggy.

"Drat the luck!" Kate slammed the screen door. "There goes the new kitchen he promised me after the crop is off."

She didn't stay mad long.

By suppertime she was calling out orders and slapping the dishes on the table at top speed.

"Get that milking done at once, or you'll not go to Kennedy with Dad and me after supper."

It was the third evening trip she had taken that week.

CHAPTER 36

Kate's first objections to a McLaughlin in place of a Ford were a sure sign to the family that their car shouted prestige. They were giddy with pride when Sam drove to Kennedy past a few lowly Fords sputtering along the trail.

School was the real testing ground of automotive supriority.

"The Baby Grande Chevrolet is better than a Grey Dorte any day!"

"Mmmmp! Our Grey Dorte can leave your Baby Grande standing still! Whatyuh say, Buster? Ain't that true?"

Buster chewed on a piece of grass and looked judicial. The "country kids" sat on a shaded side of the big board fence which separated the boys' side of the schoolyard from the girls' side, and discussed cars with the lofty air of those who had just "arrived socially".

Ordinarily the country children were shyer, less carefully dressed, and arrived at school in old democrats, buggies, or astride sway-backed nags, acquiring dust and untidiness on the way. This tended to make them feel vaguely inferior to the town children, who arrived clean.

The advent of the automobile changed all that. Good crops and high prices made farming pay. Trainloads of cars arrived in all the small towns. Most townspeople on salary were in no position to buy, but the farmer, always a gambler, bet on a prosperous future. He put up his buggy and team and, accustomed to machinery, quickly snapped up and adjusted to the cars as they came in.

"The Grey Dorte's a pretty good car, I'll say that for it. It's got a good chassey —" Buster, the centre of a ring of boys, wrinkled his forehead, paused, and considered the debaters.

"There's nothing wrong with a Baby Grande neither, mind you. You'd have to get them on the road and race them. Not much to choose between them." He paused, chewed some more grass, and cogitated. The contest for superiority rested. The McLaughlin wasn't mentioned. Buster had established its superiority the week before.

The McLaughlin wasn't a month old when the Pattersons, next door, bought a Baby Grande. A carload of children, all as pleased as the Markses had been, whizzed past as often as an excuse to town could take them.

"My Dad passed your Dad on the way to town last night." Tommy Patterson leaned over in reading class and whispered to Buster. "He can beat your Dad any time, and he was only going twenty miles an hour," he taunted.

"Our car can go forty miles an hour. Dad wasn't even trying." Buster was red in the face.

"You two boys, Buster and Tommy, stay after school and you can discuss your problem then." The teacher, gimlet-eyed and impatient, rapped two pairs of knuckles sharply, which added fuel to the fire. They were detained for fifteen minutes, given a lecture on talking in school, and dismissed. Wordlessly they bounded down the steps and raced to the schoolbarn where the other children were waiting. A circle was formed and the fight was on.

"Hit 'im, Buster! You can beat him!" shouted Buster's friends.

"Knock his block off, Tommy. He ain't bigger than you!" shouted his enemies.

"Atta boy! Hit him!" and the two gladiators swung and punched and kicked. The fight didn't last long. Buster had the advantage of a few months in age and a belief in his own invincibility.

"Say the McLaughlin's bettern the Baby Grande. Gwan! Say it!" he roared. He was astride Tommy's back with a hammerlock hold on one arm, and he pushed Tommy's face into the manurey bedding-straw. His own nose bled over his shirt front, but he didn't stop to wipe it.

The source and reason for the noise alerted the teacher, and she hurried out to stop the fight. No one in school cared to question the decision. Buster won the car argument, but Tommy still had one advantage, and he played it.

The Marks team was hitched to the democrat, ready to take off. The Patterson cart with their tough little pony, Punch, between the shafts, stood by waiting. The moment Buster whipped up his team Tommy, with his two sisters beside him, let out a war-whoop and the race was on.

The Marks team, semi-work horses with no race-track ambitions, had a load of children and a heavy democrat to pull, but they were game. The Patterson cart hit the ditch, then spun along onto the grass at the side as the team, with Buster lacing them angrily, pounded along the main trail. The children bounced and yelled as the democrat careened from side to side, hitting gopher holes, ruts and stones with equal abandon. Books and lard-pail lunch-cans flew into the air, scattering papers, lids and left-over crusts along the way. Buster steered the team from the road and across the ditch, trying to cut Tommy's pony off by sheer weight and volume. Tommy stood up in the cart, tight curly hair on end, with one black eye almost closed. He shouted and laughed and laced his sturdy little pony with a switch as it pounded along, never changing its determinded stride.

"Mooley cows! Mooley cows!
Put 'em back behind the plows," hooted Tommy.

His two sisters cowered in the cart, screamed steadily, and hung onto the sides.

"Get your mooley cows off the road!" he taunted as Punch pulled the cart across the ditch and onto the trail in front, with no apparent effort. The team lumbered along behind, foam flying from their bits and their sides heaving and lathered in sweat. Tommy turned around, laughing and thumbing his nose. Buster's neck was swollen and his face scarlet with anger.

"I'll get yuh tomorra! I'll bust yer thick head fer yuh!" he promised, his threats lost on the wind.

Punch settled into a comfortable trot, setting the pace all the way home.

Wise in the ways of defeat, Buster didn't mention the race the next day at school. Tommy, just as wise, didn't either. He had conceded defeat on the car issue, but every night after school he hitched up his pony, waited until Buster had the team ready, then guided Punch out in front. The little pony took and kept precedence over the team from then on. It was galling to Buster, but there was no way of getting a team of racers to beat Tommy.

CHAPTER 37

One delicious excitement followed another after the advent of the McLaughlin. Grandpa Marks, as half-owner of the car, demanded to be taken to Fish Lake in the Moose Mountains, a series of little, wooded hills dotted with tiny lakes, twenty-five miles to the south. He roped a canvas tent to one running-board and a roll of bed-clothes to the other. A brown tin trunk was filled with food, and it rested on the leather cushions of the back seat.

"I be goin' fishin'. 'Ee can bring the children down on Saturday." They trundled off to the lake, where Sam left him. He set up his tent overlooking the water, fished and rested in grateful solitude for a week.

For the children, the days were achingly long. Saturday came at last, and the morning sun crept up the sky with reassuring brilliance. Kate, as usual, dug in her heels which made the trip twice as exciting when accomplished.

"It's been so hot, we'll get a thunderstorm out of this, I'll be bound.

Sam, what are the roads like in to that lake? If it rains, we'll never get the car out. Mark my words!"

Sam murmured something unintelligible through a froth of shaving soap he was lathering. He stropped his straight-edged razor and carefully shaved around his moustache, stopping from time to time to urge everyone to greater speed, as if he too was afraid of Kate and the weather.

"Get a move on! If you're not ready, I'll drive off without you." He changed into his best black suit, fastened a stiff white collar to his shirt, clipped on a tie, and put on a black homberg hat.

"It's not gonna rain! It CAN'T rain today!" The children, usually happy as new ducks about showers and rivulets of water they could puddle in, watched skyward fearfully, each dumpling cloud an enemy.

"I'm going for the car," shouted Sam. "Anyone not ready stays behind." Edith tagged after him to the car shed, clad in snowy white. She wore an eyelet embroidered dress, stiff with starch, waistline bottom-high, long white stockings and white, rubber-soled slippers. Kate vainly expected the outfit would remain clean for at least an hour. The child clung to a faded blue cotton bathing suit, two sizes too big, and raced to get into the front seat ahead of the others.

Blossom, a small brown pet hen, was there before her. She sat contentedly in the hollow of the driver's seat, feathers fluffed and busily producing her daily output, one fresh brown egg. Job completed, she "cut-a-cutted" as Edith dropped the bathing suit in the car and lifted her out. Sam checked the tires and got out the crank. Edith picked up the egg and wondered what to do with it. If she took it to the house, someone would be sure to grab the front seat. If she threw it out, she'd be left behind for wasting.

"Advance the spark, advance the spark," her father shouted, and he spun the crank in front. The little girl slid onto the driver's side and shoved the lever affixed to the steering wheel to the top of a steel semi-circle. The engine coughed, gasped, and then churned into a roar.

"Retard the gas, retard the gas," her father yelled impatiently. He let go the crank and rushed to the driver's seat as Edith pulled the lever down to zero again. The engine quit, and Sam looked mad.

"Don't shove it down so fast. Move it slow or you'll kill the engine!" He ran around to the front and spun the crank again. Edith followed directions, gripped the spark with one hand and held the egg in the other. The car burst into life. The egg burst down the front of her clean white dress.

"I've a good mind to leave you home. Can't you keep clean for five minutes?" Kate scoured off the hands and dress. Buster usurped the place beside his father. Edith, with a large, wet, unstarched patch on her soiled white dress, crawled in the back with the other children.

Kate pinned a wide yellow leghorn hat, trimmed with an enormous blue ostrich plume, to her hair, wrapped the baby in a shawl and got in the front beside Buster. Her white crepe-de-chine blouse, one of the new Grandma's discards, was topped by a bib of home-made lace. She wore a long, flowing blue serge skirt, and a navy blue taffeta coat that rustled.

The car lurched down the hill, picked up speed and spun along the road west. Kate hung onto the baby and clung frantically to the side of the seat. The feathery fronds of her hat sailed out behind her like curls of bright blue smoke, furling and unfurling in the wind.

"You're going too fast, Sam. We can't see the scenery!"

"Go faster, Dad! Go Faster!" shouted the children from the back.

"We're going twenty miles an hour by the speedometer," yelled Buster from the front.

The car roared through Langbank scattering Upton's chickens and McClements' dogs, squawking and howling, from the roadway, and leaving clouds of dust behind.

The car rumbled west across the fields to Kennedy, then wound south-east towards the "mountain". The ungraded wagon trail, peppered with stones, was unprepared for cars or Sam's driving. The front tire hit an embedded rock and an explosion like the crack of a gun announced the first flat tire. Kate screamed and the car slewed sideways.

Everyone piled out and sat around on the grass. Sam took off his jacket and hauled out the jack and the mending kit from under the back seat. With the boys' help, he pumped the car up on the jack. An hour crawled by while the nauseous glue that held the patch on the torn tube dried, and the tire was inflated again. All hopped in, the car started, and they were off. The trail opened onto a graded road for the last five miles in to the lake.

"We're going to the lake! We're going to the lake!" shouted the children. Their world of flat dry fields, dotted with clumps of stunted trees, faded, and they entered a magic land where dainty birches fluttered their leaves over pools of bright water. The trail wound through aspen groves and around sky blue sloughs where clouds of red-winged blackbirds sang and swung on the wild marsh reeds and wild ducks, flotillas of young in their wake, threaded the surface of the waters.

A fresh smell of musk and a silver glitter through the trees and they were there. Harnessed horses, tethered to the wheels of wagons, munched grass beneath the trees. Families sitting on blankets eating nearby looked annoyed as the car jounced past.

It trundled along the shore trail that skirted the lake, a wonder of water that looked ocean-big. The lake was almost a mile across. Up a hill and round a turn their Grandfather's camp sat in a clearing, with hazelnut scrub and chokecherry bushes shielding it from the road.

Everyone piled out and assailed the camp like a scourge of ants, examining everything.

"Kate, 'ee lie down in the tent and I'll make 'ee some tea and a little some'at to eat." Kate went in, put the baby down on a straw tick that lay on a bed of leafy boughs, and took off her hat. Grandpa stirred up the campfire which burned in the centre of a ring of stones, and set a blackened tin of water on the tripod above to heat. Sam found a crony in the next camp and discussed crops. It was a day of dazzling discovery. The children found springs of water that bubbled out of the ground without the aid of a pump.

They tore off their clothes, jumped into stringy cotton swim suits, and waded timorously into the lake. Shouting courage, they flopped heads down, gasping and splashing gloriously. Later, shaking with cold, they staggered out, water-logged, wet suits sticking to them with embarrassing intimacy, and came to lie and bake in the sun on the hot sand.

They caught and imprisoned in pails, soft green frogs that lillipped and croaked along the shore, and used them for bait when their grandfather took them fishing. They watched the sickening death throes of the frogs when they were impaled on hooks.

"They be no good when they stop kicken'," Grandpa Marks explained, removing dead frogs from the lines and tossing their limp bodies into the lake, white bellies up and guts trailing.

Silvery pike and pickerel flipped and flopped alarmingly on the boat bottom and were later fried a crusty brown in a buttered pan over the smoking campfire. With Kate's fresh baked bread, they were the fare of kings.

The children raced back to the beach and paddled impatiently along the shore until the required hour had passed when they could dog-paddle swim once more.

After hours of rest in the tent, Kate sat up and began to think of the trip home. Outside all was still. Not a leaf stirred. The low mutter of thunder and a sudden darkening of the sky startled her into action.

"Father," she called to Grandpa, asleep against the bole of a tree, "tell Sam to round up the children. We'll have to hurry if we want to get out before the storm breaks!"

It was too late.

Moments later Sam was herding everyone out of the water and up the earthen steps gouged out of the bank, counting noses as they passed. Shivering, wet, and blue with cold, the children clutched sopping towels and pushed into the tent to dress. The men squeezed in after them.

The first drops of rain struck the tent top, then the heavens opened and dumped down a deluge. Wind exploded over the tree tops and limbs above the tent thrashed and danced and splintered, tossing down branches with a deafening clatter. Sheets of lightning, followed by sky-splitting thunderclaps, rolled across the sky. The tent sides snapped and billowed, the pegs the old man had pounded in to hold it, miraculously kept it on the ground.

"Don't touch the sides of the tent or it'll leak," warned Sam. Only a few inquisitive fingers stole out to experiment. The rest were too intent on crowding against the security of the grown-ups in the centre.

A momentary pause in the fury of the storm, and then a rattle-like machine gun fire on the car hood and a heavy thudding that tore at the top of the shelter began. Hail stones big as golf balls rained down, peppering the ground and stripping the trees of leaves. Some of the children peeked out through the slit in the tent flap. Chains of lightning flickered across the lashing water of the lake, like fiery white snakes bent on victims to attack.

"All the crops," whimpered Kate, "I knew we shouldn't have come."

"Go on, Kate," her husband bluffed, "being there wouldn't stop the storm. It's likely only local and we're a long way from home here!"

"That's just it. How will we get home!"

"It'll soon be dry. Don't worry, we'll make it!"

The steaming dampness within the crowded tent was no dampener of spirits, except for Kate who was pale with worry, and the baby, who howled a lot.

In a few minutes the wind lessened, the rain and the hail stopped, and the storm lumbered off to the east. Outside, the long rays of a late sun came out and glinted on dripping trees. The campfire was a sodden mass; tin plates and pans floated in pools where the family had picnicked, and the ground was littered with melting balls of ice.

Standing up, they ate chunks of cold fried fish with bread and peanut butter. The tent was taken down and packed bulky wet on one side of the car, the empty trunk and bedclothes roped to the other side. Everyone clambered over the load into the seats.

The car slithered down the hill and along the shore towards the road home. A cold wind, aftermath of the storm, ruffled the darkling surface of the water. A full moon, just rising, sent shudders of silver ripples towards them. Loons called lonesomely from the marshes, and a cool summer night closed in.

In hot weather the fine black soil rose in blinding clouds after every passing rig, and settled on the roads inches thick. Mixed with rain, it formed a glutinous gumbo pudding, slippery as glass, that required skill and split-second timing to navigate. Sam had both.

Homberg hat atilt on the back of his head, black eyes intent on the challenge ahead, and a certain twitch to his moustache that hid a half-smile, he drove with the abandoned joy of his race-track days when he had urged his horses to hard-won victory.

The car danced and swivelled and spun back and forth over the slimy grade. Sheets of water or a sudden wall of trees showed in the car lights and brought gasps and small noises of alarm from Kate. The smaller children slept draped over their Grandpa in the back. The older ones thrilled to the unusual contest, with complete trust in their father.

"Dad! Dad! There's a buggy up ahead," Willie shouted from the back seat. Picnickers had long since stabled their horses for the night, or left for home the moment the rain ceased. Oblivious to the dark or the lateness of the hour, two heads close together in a buggy appeared briefly in the car lights at the top of the next rise, their horse plodding along through the mud in the centre of the grade.

Sam's attention wavered for a moment, the car turned sideways and slid down the hill, and Kate screamed. Near the bottom the machine turned a half circle, swerved and headed for the ditch. The front wheels sank into the rushes and mud and the engine died.

"Everybody out!" shouted Sam. All set to work except Kate and the baby. Grandpa cut leafy branches from nearby trees to put under the tires. The children all set out to push.

The moon was well up in the night sky when the McLaughlin chugged up the next hill. Summer clothes, well dirtied before the family left the lake, were now splattered with mud and chewed-up grass which the spinning wheels had flung out.

The moon paled and the first yellow rays of the sun streaked the eastern sky when they drove into their yard. Sam lit the fire and his father filled the boiler and put it on the stove to heat.

"Get those muddy clothes off, and you'll all have a bath before you go to bed," Kate instructed. She rattled up the stove and put the porridge on to cook.

The children were all too excited over the night's adventures to think of sleep.

"I wasn't a bit scared!"

"Neither was I! I bet Dad's the best driver in the whole world!"

"Only Mama was frightened, but she had the baby!"

Kate stopped her hurrying.

"Frightened nothing! I enjoyed every minute of it. Now just you hurry up and get out of those clothes and get at the milking."

CHAPTER 38

A year went by, and Amy's education began.

Miss Murchison, the new teacher, entered the schoolroom on the first day with all the confidence of a cornered rabbit in a pitful of snakes. She wore a pretty white sheer blouse, frilled at collar and cuffs, and showing just the proper amount of lace camisole beneath. A long brown skirt fitted her trim figure. There the pleasant appearance ended. Her mousy hair had a dank, oily appearance, like wet weathered hay. Big flakes of dandruff clung to the damp strands, and the brown bone hairpins she used to keep it in place were always falling out on the floor.

Her pale eyes continually blinked. They were scaly and red from the tears she probably shed at night. Sallow skin, pocked with tiny holes, and a facial tic which pulled one corner of her mouth up in a flickering grin and which she nervously tried to hide with her hand, added nothing to her charm. Besides these blemishes, her worst feature was an enormous outsize nose.

She had a language tic, too, which the children, flattered by her fear, grabbed onto like hounds at the heels of a frightened deer.

"Turn to page 65, grade 8," she mumbled, "and commence your reading." She looked perturbed, as if she expected resistance.

"Commence your sums, grade 3." "We will commence the Family Compact, grade 6." "Recess will commence at ten-thirty." "Commence — commence — commence."

Eleanor, Hazel's seatmate, looked up from her work. She stared at the teacher, then she turned to the back page of her exercise book and printed "COMMENCE" in large letters at the top, several strokes beneath. A whispered conference, and Hazel did the same. Eleanor then rose from her seat and, exercise book in hand, passed slowly down the aisle like a busy teacher's helper, to explain her new strategy to her grade.

The morning progressed. Each time the word "commence" was uttered, one line of children quickly turned the books over and made another mark in their columns. Miss Murchison's eyes blinked in alarm each time the books slapped over, but she made no move to discover the cause.

"I've 67," remarked Eleanor at the end of the day. School was about to end and the children from her aisle were gathered about her desk in a tight circle.

"I've only 65."

"That's wrong! I counted 68." the argument, spoken aloud, went unnoticed in the general din.

"You missed two when you went to the toilet."

"You gave me the score when I got back."

"As far as can be ascertained," remarked the efficient Eleanor, "my score must be taken as correct," and she checked each child's book just before dismissal.

The teacher tapped the bell and the children exploded through the door and raced down the steps. They stopped abruptly, and gathered just outside the gate, by an old implement shed. Its wall, a perfect place for chalked graffiti, sported a larger than life-size profile of the teacher. The bun of hair at the back of the long neck, and the gross oversize nose, were unmistakable. To make doubly sure, the artist had scrawled "TECHUR" beneath it.

"Who did it?" quavered the subject of the caricature.

"I seen Tommy go to the backhouse last. I think he done it," Albert, perpetrator of most jokes, supplied.

"I think it looks more likely like "Billy's drawing," Tommy put in quickly.

"T'wasn't me!" Billie, always reasonable, answered. "Girls are good drawers. Which one has chalk on her hands?"

Everyone crowding around the teacher quickly wiped hands on pinnies or pants and innocently viewed the offending picture. The discussion, unsolved, came to an end when the girls, headed by Eleanor, tried to scrub it off with water and the school brooms.

From then on every wall, board fence or blackboard flaunted the caricature — large nose, sagging hair line and mournful mien, of the unfortunate woman. Like ants at a picnic, the cartoons turned up with startling frequency. Hours of school were joyously syphoned from arithmetic and other unpleasantries to find out "who done it," but only one culprit was ever caught.

Amy, a beginner, determined she could do what the elder children did, quickly learned to read.

"I dowanna play pum-pum-pullaway! I wanna do adds!" she decided one noon. "You help me, Edie." They got behind the organ leaning against the blackboard, and went to work.

Teaching lost its interest. Edith wandered off, and Amy began to draw, like everyone else. Deep in creativity, she finished to satisfaction the outsize nose, the sagging hair, the long neck, when a clearing of throat sounded above her.

"Amy! What ARE you doing?" Indignation and disbelief in every tone, Miss Murchison's voice rose in a kind of bleat as she peered into the dark recesses behind the organ.

From that moment Amy set a precedent for quick thinking and how to turn a catastophe into a triumph. With the same aplomb she used down the years, she said, "Oh, Miss Murchison! I'm drawing your picture but I can't do it very well. You're too beautiful!"

Her trembling voice and the tears about to spill out of big brown eyes touched the frantic teacher. Miss Murchison suddenly burst out laughing, gave Amy a slap on the bottom, and shooed her out of doors. From then on interest waned in drawing. November came with snow, and the school settled into winter programs.

CHAPTER 39

The biggest event of the school year was the Christmas concert. The children worked up to concert time by having Friday afternoon programs when everyone had a chance to "do a piece."

"Would you do a recitation?" Eleanor enquired of Edith, pencil poised over a list headed "LITERARY SOCIETY — PRESIDENT — ELEANOR." "Or a song perhaps?" She leafed through her notes.

"Here, I've just the thing. I'll play the organ and practice you at noons. I'm sure you'll be a success."

Eleanor thumped the organ and Edith, overcome by the honor, quavered through this song:

"My mother told me that she would buy me
A rubber dolly if I'd be good!
But don't you tell her I've got a feller
Or she won't buy me that rubber dolly."

Friday noon came.

"I won't be playing for you, Edith, for of course you sing without accompaniment." Eleanor made this surprising statement, shuffled her notes busily, and trotted off to check on her other performers.

The bell rang and the concert began. When her turn came, Edith was overwhelmed. She stood centre stage, all eyes riveted on her. The teacher smiled encouragement as she began. Before the song was finished the teacher's attention had wandered and she stared unhappily out of the window.

"You'll have to pick better material," whispered Eleanor severely. "Obviously the teacher does not approve of reference to boys at your age!" She sailed off, leaving Edith crushed with embarrassment.

"Couldn't you do a recitation this Friday?" enquired the president some weeks later. "A more suitable one this time — something sad, perhaps?"

Exhilarated by this second chance at redemption, Edith chose "The Face at the Window," a piece about a lonely child, funerals all over the place, woe in every verse and tears dripping from every line. A black wreath hung on the door, starvation and abandonment hung over the child's head — truly a recitaion to soften the stoniest heart.

Visons of the whole classroom quietly weeping as she ended her elocution, a standing ovation perhaps, spurred Edith on. She would simply bow at the finish and take her seat with humble dignity.

The occasion had lots of drama but no pathos.

On Friday afternoon she took her place at the front, cleared her throat for attention, and began.

"Just next door there is a stranger," she intoned, "and her face is wet with tears —" she paused for dramatic punch, got it, and took a deep breath as all eyes riveted on her.

The quiet was suddenly shattered by the greatest of all school embarrassments, the dreadful bathroom expulsion of air. It reverberated from the back of the room like the loud blurt of a locomotive, followed by several short, sharper blurts from the same source. A shocked silence was followed by a thunderous explosion of laughter.

"Stop that noise at once!" Miss Murchison's eyes blinked rapidly and her hand fumbled towards her mouth. "Who made that noise?"

A dozen hands shot up.

"I think it came from that corner over where Billy is sitting!" Albert volunteered.

"Albert's mistaken. I'm sure it came from over there where Tommy's sitting."

"No, Miss Murchison, it sure sounded to me as if it was around Buster there . . ."

Edith stood paralyzed at the front of the class, her moment of glory turned to a nightmare of ribaldry by these monstrous classmates who argued, debated and talked among themselves until the clock struck four. Everyone then arose, and sighing with mock concern over the undiscovered culprit who made bathroom noises in school, raced outside to laugh all the way home. They had no regret for having missed the sad and stirring epic of "The Face at the Window."

CHAPTER 40

It was June. The trees to the northwest were a joyous green, clear as emeralds under water. In the north field, a man was discing. His four horse team arched their necks against the pull of the heavy machine behind them. They planted their feathered feet rhythmically in the soft soil, churning up dry prairie dust that trailed in clouds back and forth across the warmed spring fields.

To the west, a moving speck appeared at the top of the hill. It was the blunt black snout of the McLaughlin, rapidly chugging toward home. Sam was bringing the Winnipeg Free Press and the Saturday funny papers. Waiting for the funny papers was a weekly anticipation, equalled in intensity to the expected arrival of the Eaton's clothes parcel or the waiting to get paid for gopher tails.

Sam brought in the papers, passed the funnies to the children, and sat down to read bits of war news out loud.

"Listen to this, Missus!" he began. War news, lists of the wounded or dead, were always weeks old by the time he read it in the papers.

"Ervine Johnston wounded; Herman Johnston killed in action; Tom Cairns killed in action; and Wilbur Potter wounded." All these local boys had been on the black-edged casualty lists over the months since the war began.

Kate listened as she rushed from kitchen to pantry to dining room, stepping over the children who were arguing over the funny papers spread out on the dining room floor.

"I'll trade you Buster Brown for Mutt and Jeff!"

"Gimme Happy Hooligan, then."

"I just love Gaston and Alphonse." This from Hazel, who was always impressed by the good manners of her two heroes.

"I like Hans and Fritz the best." Amy pored over the "Katzenjammer Kids" while Edith read out the captions. Their favourites were two little boys who planned and executed tricks played on their fat father, "Der Captain," and his mangy little friend, "Der Professeur."

"They're Germans!" scoffed Buster.

"No, they're not. They're Belgians!" The children felt in their hearts this was not true, but German or not, they loved their defiance of all things adult.

The papers were jammed with hate propaganda of the dreaded "Hun." Pages were devoted to drawings of horror, burning houses, dead babies lying in the streets, and helmeted soldiers goose-stepping over them. A cat nailed to a door instilled terror of the enemy in all the children.

"I hate those Germans! Look what they did to that poor little kitty!"

Following Sam's example, Edith read out loud every chance she got.

"You can help." This terse message was at the bottom of a big picture of George V. Another full page was taken up with the martyred Kitchener who had drowned, a war hero, in the North Sea. Black-browed with bristling moustaches, he stared out like his ghost from his watery grave. He pointed a finger straight out and the caption beneath said: "I NEED YOU" which Edith read loudly.

"Never mind Kitchener and the King. Right now I need you!" Kate grabbed the scattered papers from the floor and pushed her daughter towards the wash basin.

"You wash your face and hands and go to get a clean flour bag off the line. Buster, you hitch up the team to the democrat. You two are going to Boyd's for the meat."

The family belonged to the meat ring, an association of farmers who supplied one steer each during the summer months. Mr. Boyd did the butchering. In those iceless, freezerless days, members were assured of fresh meat at least once a week.

As soon as Buster was ready, Edith climbed up on the democrat seat beside him, and they set off east towards the old Indian trail that branched to the north across the prairie still unplowed. The trail meandered around knolls, over little hills, and skirted tiny sloughs where the grass grew tall and the bullrushes swayed in the wind. Killdeers piped and hopped on their thin legs, making thread-like tracks in the mud. Wild ducks nested near the water in the grass. Meadowlarks, black v's shining on their yellow breasts, sang their cheery songs, and crows flew back and forth above their heads cawing their questioning cries.

The harness jingled pleasantly, the horses snorted and coughed as they clopped along, and the buggy wheels hummed as they spun over the soft black earth.

Suddenly the deeply worn ruts seemed to come alive.

"Buster! Buster!" screamed Edith. "Look! Snakes! Snakes!" and she grabbed his arm.

Orange, yellow and black striped bodies shot like projectiles across the path, the bright colours in startled contrast to the grey-green prairie grass, beguiling the eye and tingling the spine with unnamed terror.

Suddenly they weren't there. Only the grass disturbed further on proved eyes were not betrayed.

"Aw, don't be silly. They won't hurt you!" Buster clucked to the horses with the superior air that only an eight-going-on-nine year old boy can attain when doing a man's work. They sifted along the dusty ruts, the wheels making scritching noises as they bumped over stones that stuck up in the winding trail.

"I wish I knew how we could help Kitchener and the King." Edith's mind was still on the propaganda demands in the paper as they jogged along. War atrocities, gleaned from adult conversation and lurid pictures in the papers, had peopled her constant nightmares with vicious soldiers bayoneting everyone in sight.

"Where could we hide if they got here?"

"I've got a good safe place!" Buster clucked at the horses again, spat over the wheel, and stalled for time while he thought one up.

"I'd hide in the hay in the loft and cover all up. They wouldn't know I was there.

"Awww! don't be silly. You wouldn't stand a chance. That's the first place they'd look. They'd stick bayonets through the hay and there you'd be!"

"I wouldn't say a word if a bayonet stuck through me!" Edith was determined to be heroic.

"Awww, g'wan. You'd yell first thing!"

Edith wisely waited for his solution.

"Git up udavit!" He slapped the reins on the horses' backs and spat again, through his teeth this time like the hired man did. "Git up udavit!" and the horses, sweat beginning to shine on their wine-coloured sides, lurched from a slow trot to a trot.

"I've got a place!" he announced at last.

"Where? Tell me! Could I come with you?" Being left behind to bleed all over the hay wasn't pleasant.

"You're not to tell if I tell yuh. I've got it all figured out. The gopher place is one place they wouldn't bother to look." The gopher place was a few acres enclosed behind the barn for evening grazing. It was pocked with gopher holes.

"I'm gonna dig a hole just over the hill that runs along the ravine there. I'll put food and water and everything I need in it in case they stayed around. I'd have a cover made of sods so it would look just like the rest of the prairie. I'd spy on them from my lookout till they went on to Langbank!"

"Couldn't I come with you?"

"Nawww — well, I dunno. If you could keep quiet. Naww — you'd tell or something. Then there's the diggin'——"His face brightened, and he relented somewhat. "I guess, maybe —" His ideas usually found her willingly carrying, digging or pushing something while he directed.

"Well, maybe — but you'd have to be awful secret about it — not tell anybody, specially in the family. There's always spies around, you know."

"I wouldn't say a word to anyone! Honest! What about the other kids? Would we make it big enough to hold them too? We couldn't leave the baby."

"There you go. How could we hide ten kids in there? Too many tracks and the Germans would be sure to notice. They're smart, y'know. Nope. We've got to figgur someone's got to die. Make up your mind now". He looked grimly adamant, so she was forced to concede.

"I'll dig and I won't say a word, honest, I promise." Edith privately vowed she would dig twice as hard and make it big enough to include at least the baby.

Kitchener's accusing finger still burdened her conscience.

"Mama says not using sugar in your tea helps the war effort, but I can't see how." They had just learned to drink sugarless tea with little regard to taste, but it was enjoyed as a strange adult pleasure — a poor defence against an army of soldiers. "I wish we knew of some other ways. . ."

"Wanna know how? There's what I bin lookin' for!" Buster stopped the team, handed her the lines and hopped down over the wheel.

A group of balloon-like mushroom growths dotted the prairie. He kicked out as many as he could find and sauntered back to the democrat.

"The hired man says those are the Kaiser's balls," he explained. Anything smacking of the Kaiser's possessions was worth destroying. They tied the lines to the dashboard for the next ones, and they both jumped out and kicked enthusiastically.

"You're late, children." A tall young woman, cornflower-blue eyes, freckle-spattered nose, and a pile of burnished hair on her pretty head, tripped down the path between the birch trees towards the meat-house where they had stopped the team.

"Better go up to the house for some cookies and milk." She put a great lump of bleeding meat into the flour bag and lifted it up for them.

They clambered down and sidled towards the house. The aroma of Saturday baking, fresh bread and oatmeal cookies, greeted them at the screen door. In the Boyd kitchen several young people surrounded two handsome soldiers. They were all talking and laughing together.

The children munched cookies and sipped milk and looked out from under their thatches of windblown hair at the first two uniformed soldiers they had seen in the flesh. Both were tall, with sandy gold hair, brushed and shining. Wide leather belts encircled slim waists and across chests. Their trousers were "breeks" of a lighter colour, and stuck out at the sides. High leather puttees encased slim long legs, and they tapped the leather from time to time with short canes they carried. Their faces were open and smiling.

Edith's hopes for the future swung from a dirt hole behind the barn and came to rest in the hands of these two handsome soldiers, Bill and Jim Boyd. No German would dare try to get by them.

"I'm going to marry one of those Boyd men when I grow up, but I don't know which one I like the best," she said dreamily back in the democrat.

"Yuh'd better wait to see which one will ask you," Buster advised.

They dawdled all the way home. They killed a snake and stopped frequently to kick out the rest of the mushrooms missed on the way there.

"What kept you so long?" Kate was impatiently waiting to put the meat on for supper when they drove into the yard. "I phoned the Boyd's and they said you didn't get there till way past noon."

Edith jumped from the democrat, full of enthusiasm.

"We stopped to kick out the Kaiser's balls! There's not one left between here and Boyd's. We kicked out every one!"

"What?" Kate shouted, "you did WHAT?"

"We were helping the war effort like the King wants us. If the Kaiser comes and can't find any of his balls he'll sure be mad."

Kate assured them that that wasn't the scientific name for the mushroom, but they never quite believed her. They secretly destroyed every growth they found, feeling it was one way to confound the enemy.

The dugout was started in the gopher place that evening, but after crops of blisters Buster wisely decided to put his faith solidly behind the Boyd men too, and the project on the gopher place was abandoned.

Violet Alice was born that September, and the end of the war came in November.

The morning the news that the war was over was phoned in from the Central Office, everyone shouted and sang. Willie played "Casey Jones" on the piano, loudly and triumphantly.

CHAPTER 41

There was no dearth of subjects to be discussed with visitors that next year. The Turk's cruelty to the "Starving Armenians" filled the papers and took over the children's nightmares where the goose-stepping armies of vanquished Germany left off. Alsace-Lorraine and its restoration to France was headlined for months. Reports of a dreaded malady sweeping Europe, supposedly started in Spain and called "influenza", appeared with fearful frequency in the news.

The "boys" that were left began to come home from the front.

"What was the good of all those lives lost?" Kate seemed more bitter as the year drew towards the end. She was weary and short-tempered after the threshing was over.

"No use to think about it, Kate." Sam paused as he searched the papers he had been reading to her as she worked. The events in Europe were still important to the prairie farmer.

"Too bad they hadn't put the Kaiser and the King in a ring at the start and let them go to it. Saved a lot of good lives; Tom Cairns, Herman Johnston —" He sighed, took off his glasses, and folded the paper. "At least it would have got rid of a couple of people who ain't much use anyway."

> "Then the soldier, full of strange oaths and bearded like a bard, sudden and quick in quarrel, seeking the bubble reputation even in the cannon's mouth ——".
>
> Shakespeare.

Shakespeare's lusty young soldier was a far different thing from the sad young men who limped back into the district, some lacking an arm, some a leg, and others with deeper scars.

Johnnie Binder, a quiet young veteran, employed by a lumber company, came to town with his pretty English bride. He was welcomed by all the farmers anxious to build new pens, granaries and barns with money left from farm profits due to the war.

In the middle of a wild thunderstorm, Binder went beserk, got a gun he'd brought home, and began shooting wildly, narrowly missing his terrified wife. He was taken away quietly the next day. His wife left too, and nothing remained except his memory, long critical discussions as to why he didn't settle down, and the bullet holes in the ceiling of the little house in which he had tried to live for so short a time.

The returning soldier, plagued by his hideous dreams, was accustomed to the beguiling powers of the bottle, which the happy unpressured people left at home had long since ostracized as being an unnecessary evil in the community. Unable to buy branded stock, the veteran made his own or bought it from someone else who did.

Sober, six-foot-two "Wee" Willie McCuaig came home, went back to work, and began to court the new teacher. To the district he appeared to be the same sober, shy lad who went away, unchanged by the deadly four years he had spent as a machine gunner at the front.

The night of the Christmas concert came, in which the teacher played the major part. She had trained, accompanied and chaired the Programme of her charges to be performed before her main critics, the parents. In the middle of the program "Wee" Willie, his heart full of love and his belly full of home brew, could see the teacher needed some help. She went behind the drawn curtain to assemble the children for the next act, and "Wee" Willie staggered to the front.

He climbed on the platform, turned to the audience, bowed from the waist, and bellowed to the astonished gathering. "Ladies and gentlemen, I come tonight to address you, not to undress you — not to undress you — not to undress you —"

A wave of laughter from the audience, and Willie's face changed. "Die, you sons of huns! Die!" and in the stance of a machine gunner, he swung from side to side, in imagination wiping out the faces in front of him.

Two men leaped to take hold, and led him away. Willie took the next train west. People lumped his spectacular address with the "shell-shock", a condition that covered those mental scars which appeared so often in all the returned men.

Winter arrived early in November of 1919. On the morning of the sixth, a strong north wind, fore-runner of a blizzard, blew skifts of snow which moved and danced across the surface of the iron-hard earth.

A neighbour woman came in hurriedly, stamped the snow from her boots, took off her winter things, and disappeared upstairs to Kate's room. A short time later Dr. Hilts arrived, dropped his winter overcoat, crisply ordered some water to be boiled, and disappeared upstairs too.

Ruth supervised the dishes and put the bread into the oven to bake. The younger children, strangely uneasy, hung around the stove, whined, quarreled, and waited. The tension ended when the doctor came downstairs with a baby girl, a round, placid ball of a baby. She had long black eyelashes that swept her cheeks, and didn't cry much.

"Listen to this, Kate!"Sam rushed in to the delivery room the day following, with the newspaper. "Look what's happening over there in France!" He sat down in the middle of new-baby-confusion and read aloud the latest happenings (two weeks old) in Europe. The collapse of German economy, the Ruhr Valley, and the French province of Alsace-Lorraine were of great interest.

"How's the baby?" he asked, after all the news had been exhausted and they had discussed world affairs.

"What will we call her, Sam?" Kate plucked at the bed clothes as if one more name was just too much for her. Sam's attention was back on a cartoon of a beautiful woman representing Alsace-Lorraine being returned to her rightful place after being imprisoned by the cruel enemy, Germany.

"Why don't we call her Alice Lorraine?" He showed Kate the cartoon. "Kind of historical. We could call her Alice for short."

"Sam!" Kate looked aghast. "That's what we called the last one! We can't have two Alices in the family!"

"The last one? Who's that?" Then he focused in on the problem. "That's Violet, isn't it?"

"We named her Violet Alice! For heaven sake, Sam. Can't you remember your own children?"

"A few more and we won't be able to find names for them!" and they both began to laugh.

"Well, let's settle for Lorraine, then." And so they did.

CHAPTER 42

As winter progressed, news of the dreaded influenza, as it swept towards Saskatchewan from the east, was constant anxiety. After Christmas, Mary and Edith went up to Will's with their Grandpa to spend a few days and "see the New Year in".

As the old year died the fire crackled in the big round heater, throwing strange shadow pictures across the plastered walls. The wind howled and whined around the corners of the house outside, and the girls cuddled down in a bunk by the wall and listened to the grownup conversation.

Grandpa stood up to put a disk on the gramophone and wind it up.

"Play 'My Old Kentucky Home', Grandpa," they urged.

Eerie delightful trails of music wavered from the horn. The song wheezed out in a mournful voice. The music ran down and stopped. Frank

Swain, Will's hired man, said, "The Turks are far more cruel than the Germans — look what they are doing to the Starving Armenians!" There were those Starving Armenians again, who seemed to be standing around for months without food, while the cruel Turks heaped all manner of cruelties upon them.

"They be sayin' more people be dead already in France from this 'ere infloohenza than from they Germans." Grandpa stoked the fire and relit his pipe.

"It'll be comin' 'ere fast, I'll be bound, and 'll be takin' a good few with it 'fore spring!"

His prediction came true with a vengeance.

January brought increasing news of sickness from the "influenza" as the dreaded disease crept nearer across the frozen plains. A blinding blizzard that filled the road allowances fence-post high with snow raged across the land, cutting off all communication with the outside world except through the telephone.

After some deliberation, Grandpa Marks phoned Sam.

"Mary can go to school up 'ere, Sam," he announced. "Edie can go too — save 'ee some, too, Sam, if they be 'ere. They can drive my 'orse." Arrangements were made with the Poplar Grove teacher for her assistance, guidance for Mary's first year in high school. Edith was to go along to add a few gems of learning to an already very slim store.

They started out one bitter winter morning. Grandpa hitched up fat old Darling, lined the cutter with robes, and put in a footwarmer (a small, carpet-covered tin box with hot coals inside to be put at their feet). He tucked them in and handed the reins over to Mary, bade them "Look sharp now and take courage", and they set off.

"It was very slow going, for the horse had to break trail through snow up to two feet deep. She floundered on, and by the time they reached a break in the field fence where they must take the road allowance, the girls were stiff with cold. The icy wind had long since whipped the warmth from the blankets and cooled the coals at their feet. Tears of pain from stinging hands and feet squeezed from Edith's eyes and formed ice on her scarf-covered face.

The errant wind, which usually swept the road clear, had now blown it full, so that the snow was twice as deep on the road as the way they had just passed. Mary urged the horse through the break in the fence. Darling plunged on, and after the first few steps was belly-deep and unable to move.

The snow's too deep ahead, Mary," Edith yelled, "we'll have to turn around and go back!"

Mary pulled on the reins and slapped the horse's back with a willow switch. After a few fruitless plunges which only tangled her harness, the mare stopped, and with heaving sides and head resting on the top of the snow, she waited.

"We'll have to move, or we'll freeze to death!" Mary wrapped her sister in the robes and jumped into the drifts, meaning to guide the horse's head around. She struggled to the front slowly, waist deep in snow, grabbed the bridle, and pulled.

Darling reared forward and the girl fell backwards in front of the plunging mare.

"Get out of the way of her hooves!" Edith screamed, yanking at the reins.

Mary struggled upright and grabbed the slack neck-rein again.

"Steer her across to the other field," she shouted, "we can get over the fence there." Step by step they moved and stopped — another foot and they stopped again. The cutter was now in the lee of a scrub bush with trees all around, and it suddenly seemed much warmer.

Another half hour went by until at last the winded horse stood in snow she could plunge through without stopping. Mary got into the cutter once more, they wrapped themselves with the robes, and proceeded through the fields to school. They were two hours late. The teacher, annoyed at having two new pupils from another district, listened dis-interestedly to the account of their trials through the snow.

Mary made several more attempts to continue her education, but Edith never went back to that school. The project, began with haste, was soon abandoned. More dreadful events occurred that put all thoughts of continuing education out of their minds for several months.

CHAPTER 43

Several days later the two girls, covered in robes to their necks, were seated beside their grandfather in the jumper on their way home. The "jumper" was a homemade box with seats set low, and two planks turned edgewise for runners underneath, The high sides protected them from the force of the wind, but the rough contrivance whumped from one hard frozen bump to the next and was bone-jarring and uncomfortable in the extreme.

"They all be sick with the influenza, I'm thinkin'," shouted the old man over the wind. "Better you should stay 'ere with us, but Sam says 'ee needs 'ee 'ome,"

The jumper lurched and thumped along. The horse's hooves squeaked out a frosty rhythm through the snow, and the jingling harness made pleasant music in the sharp winter air. The girls were warm and glowing with the thought of going home. They sang, "My Old Kentucky Home", Mary in her high soprano and Edith following. Grandpa made no comment, listened, and urged the horses to greater speed.

It was to be the last happy period they were to enjoy until the horror they were about to face had dimmed in memory, many months ahead.

They drove into the houseyard of home, still singing. No voice answered their "halloo", no dog barked and no trail in or footsteps out marked the smooth surface of the snow. The windows, frosted like blind eyes, were blank of family faces looking out to "see who was coming". The girls threw off the robes, jumped out and ran for the kitchen door while Grandpa tied up the team.

Sam turned from the cold cookstove where he had been slivering a block of wood for kindling to make a fire. His unshaven face was a pasty white and his big black eyes stared out at them as if he didn't know who they were. A deadly cold permeated the place. Piles of ashes were strewn about the unswept floor to cover messes made when he forgot to turn out the dogs. The dogs themselves, two little terriers and two collies, lay under the table. They looked up, surveyed their human friends, then lowered heads to paws again as if they too suffered from this strange malady which had taken over the house.

Dirty dishes were piled all over the table. In the centre of the room lay a great heap of flour-bag sheets, pillow cases and blankets, reeking of vomit and streaked with blood.

The house seemed deathly still, except for the occasional sound of whimpering and a muttered groan that floated down from above. The smell of illness was everywhere.

Grandpa Marks pushed in behind the two girls. Sam swayed on his feet, and his father rushed to grab him.

"Be 'ee sick, Sam? Be 'ee all right?"

"Better go back at once, Father. Just like a charnel house here. Missus in bed. Every one of the kids sick too — bleeding at the nose like stuck pigs. Willie and the baby worse off. 'Fraid we'll lose them both if we can't get the doctor. Mike Hayes is lookin' after the stock for me." He turned to the stove and rattled the dampers and set in the kindling.

"Bring in some wood, Father, 'for you go. Haven't had more than an hour's sleep a night for over a week. Fell asleep and the fires went out." His face was wet with tears he didn't bother to wipe away.

The old man brought in armloads of wood and left. Sam pulled a mattress into one corner of the dining room and fell asleep on it, while the girls went to work.

The floors were swept, and snow dragged in to be melted in a copper boiler, for washing, on top of the stove. Bed linen was soaked in a washtub, hand-washed, wrung out and carried outside to the line where it was frozen stiff in a few minutes. It smelt cleaner than it looked. Later it was brought in to dry and be ready to go on the beds in the morning.

Kate lay, half conscious, in the parlour off the dining room, with the new baby by her side. She woke from time to time, tried to care for the child, then fell back exhausted and slept again. Mary washed her with warmed soft water from the melted snow in the boiler, bathed and changed the baby, and proceeded to the upstairs rooms. Each bed contained the fever-wracked bodies of brothers and sisters who half-opened their eyes, murmured "water", drank from the cups that were held to their swollen lips, and closed their eyes again. The two girls bathed each as best they could, and swept the rooms.

By nightfall the house had some semblance of order. Sam had two hours of uninterrupted sleep, and they had gathered and washed enough garments for clean nightclothes to be dry by morning.

"Hello!" A knock on the door and Mike Hayes came in, stamping the snow from his boots, and set down pails of fresh water from the well. He brought in more armloads of wood and pails of milk before he left to look after his hired man who lay sick at his shack.

"Bring those robes here, Edie. We'll have to sleep on a mattress near the fire." Mary fixed a bed on the floor. They wrapped the robes about them and settled to rest. They slept fitfully, listening to the strange night sounds rattling the dampers, Sam putting more wood on the fire, someone whimpering in a room upstairs, and the sad sighing of the winter wind as it whined at the windows and banged the doors.

Edith woke late, feeling cold and stiff.

"Get the floor swept again, Edie. We'll try to wash it. And let those dogs out, too. Take some water upstairs for the girls to drink." Mary was washing more clothes and trying to dry sheets for the beds. their father staggered in with more armloads of wood.

The stairs seemed very steep and hard to climb. Edith staggered into the first room, her legs refusing to work in their usual way. Hazel and Amy lay in bed, faces scarlet with fever and both asking for a drink. As Edith bent over to help them, their faces and the floor all seemed to swim together. "Move over," she said, and settled backward into oblivion.

The days that followed were vague and painful. The girls woke from time to time to see their father, haggard and anxious bending over them. He lifted each one and tried to feed her hot chicken broth. With trembling hands he guided the cup towards each child, coaxing them to eat. On the stand beside the bed were peeled oranges and a big bowl of canned tomatoes.

"You try to eat some of these now, and some tomatoes — they'll quench your thirst." He shook his head hopelessly, sighed, and left them.

In ordinary times, porridge, potatoes, salt meat, pickles and blanc mange pudding (a white cornstarch mixture sometimes graced by a daub of jam) was staple daily fare for such a large family. A can of tomatoes thrown into a large pot of macaroni was a rare treat indeed, and the children were sometimes allowed an orange as a special treat if they were sick.

The girls, now awake, struggled to a sitting position in bed. They looked like amaciated strangers. Their thin hands picked away at the blood-smeared blankets. Their cheeks were sunken, and their eyes looked too big for their faces.

"Imagine! All that lovely food that we can have if we want, and we can't eat it!" Amy quavered their thoughts aloud. They stared at each other in wonder and then at the food, favourite treats, untouched by the bedside.

The echoing silence of the house was suddenly shattered by the sound of dishes and pans falling, and then a crashing thump. a moment of silence and then the high, unearthly sounds of their mother's screams went on and on and paralyzed them with fright.

They shoved off the blankets, half crawled and half stumbled out of the room to the banisters, and peered down through the railing to the dining room below.

Sam, his body ominously still, lay stretched face down across the filthy floor, among a scattering of broken crockery, pans and ashes.

He stirred, raised his head and stared up at the girls.

"Get back to bed!" His voice was hoarse, and his face looked black and strange. They turned and tottered back, fell into bed, and were asleep in a moment, exhausted from the only exertion they had made for days.

One day they awakened feeling light-headed, but better. The unfamiliar thump of the washing machine engine and the sound of quick, purposeful footsteps echoed through the house. They were washed and lay on clean bedclothes in a clean room. A beautiful woman, dark-eyed and stately in nurses' uniform, swished in, brushed out their lank and tangled hair, and gathered up clothing to be washed. She brought in steaming bowls of broth, saw them begin to eat, and hurried on to the next room.

"I thought I died and she was an angel," murmured Amy between sips of her soup. "She sure seems like an angel!"

Some time later the nurse's resonant, silvery voice floated up from below, as she talked with Sam and the doctor.

"A real miracle, Doctor, that Willie and the baby survived."

"We can thank you for that, Nurse Burns," the Doctor's crisp voice replied.

"I think it was the mustard plasters Sam used that really saved them!"

Unaware of the listening ears above, they discussed the deaths in the district.

"Mrs. Hamilton and her new baby died last week, buried yesterday, with nobody at the service."

Mrs. Hamilton lived a short quarter of a mile away. The little girls looked at each other and began to cry.

"Quite a few gone down at the Mountain. Funny thing — so many of them — most of the deaths were young mothers. What a loss to their families. And poor Mrs. Davis! Three people at her service. But I suppose you know about her funeral from Rev. McDougall." More remarks were made, with the strange sound of shared laughter, and the children slept again.

Mrs. Burns, the nurse, stayed two more weeks after the family was better, to help Kate feed and strengthen the children.

"How much do we owe you?" Sam asked one day.

"Wee- - -eelll, twenty-five dollars, I guess."

Sam peeled off the money and handed it to the woman who had nursed thirteen gravely ill persons back to health.

Stories were told and re-told of sorrow and privation, as each family got better and took stock of their losses.

Cousin Robert came for his usual visits and recounted many happenings, the most interesting one concerning the death of Mrs. Davis, who lived a short distance from town.

In the Davis household, five family members lay ill when the mother died. Mr. Davis sent word to Rev. McDougall that he wanted him to conduct the funeral service, and the old man arrived in town for the purpose. He was met by B. A. Cotter, who performed the duties of magistrate when the occasion arose.

"Where do you think you're going?" demanded B.A., towering over the reverend gentleman. "You're not to go to the Davis's. There's to be no funerals!"

Cousin Robert stepped back and surveyed his accoster for some moments.

"Mr. Cotter! I advise you to attend to the local court's business as well as your own. I shall attend to God's business and mine! Good day to you, sir!"

He turned his back on Cotter and walked off across the snowy fields, murmuring his proposed service to himself as he went.

The great influenza epidemic ended, and as spring approached everyone settled again into the regular channels of living.

CHAPTER 44

Harvest time came around again, and all the air was filled with stubble-dust and hope. You breathed it in the golden motes that flickered and sifted from the spumes of straw which belled out of the thudding machines across the wide flat plains. It crackled like electric jolts through all the days, and filled the crisp autumnal nights with a kind of money-conscious frenzy.

Blinding dust eddied and whirled about the shorn fields, churned by the wheels and horses and men who reduced the waving grain to wagon loads of yellow wheat. This meant lots of money — a little money — or no

money at all, depending on the unexpected; frost, hail, rust, drought, grasshoppers, God or a tumbling wheat market.

"Excursion's in!" Mr. Upton at central, the nerve centre of the rural telephone system, made the terse announcement one hot morning in August. It meant a trainload of men from eastern Canada and from every European country were pouring through, and were now available for work.

Sam oiled, greased and overhauled the threshing machine and the Titan engine that primed it. He rushed in for his morning lunch of fresh bread, jam and hot tea.

"I'm going to town to see if I can pick up some good men." He cranked up the McLaughlin. "We'll be ready to start in the morning — twelve or so extra for supper, Missus," he yelled over the backfire the car was making as it gradually worked into a business-like hum.

Kate "mmmned" in agreement, straightened up from her mound of fresh bread dough being punched for baking, and wearily shoved back a strand of her hair. Men for hire were plentiful after the war was over, but a hired girl was impossible to get in a family with so many children. Grandpa Marks and the older girls after school were Kate's only help.

That afternoon was filled with more than usual activity. Kate filled and refilled the oven. The yeasty, mouth-watering perfume of fresh-baked bread permeated the house. Pantry shelves and a long kitchen table were laden, row on row, with crusty brown loaves, glistening on top with a mixture of hot butter and sugar that she painted on a few minutes before they finished baking,

Following the bread baking, some ginger cookies, a big chocolate cake, and several pies followed each other in and out of the oven.

Grandpa worked steadily, peeling pails of freshly dug potatoes. He kept the woodbox filled, and stoked and restoked the fire.

Huge pots of plums and apples bubbled on top of the stove. The tantalizing smells attracted clouds of flies, which blackened the screen door and windows. They swarmed into the kitchen every time a door was opened, and made a constant whining hum as they hovered, waiting to pounce on the baking.

"Ruth, you pull out those fly stickers and get a chair and put three on the ceiling of each room down-stairs. Mary, you go and get a saucer and put some water on those poison pads. Put some in each room, and put them up out of the way of the baby,"

The fly stickers were long strips of sticky paper, with a tack at the top to stick in the wall. Moments after they were hung they were black and buzzing with flies. The poison pads were round disks of felt paper impregnated with poison. These, too, were soon covered with the dead and dying.

"Hold the door open, and Hazel, you grab a dishtowel. We'll try to drive some of them outside."

The towels were flapped and the door held wide open, but more flew in than flew out.

"Close the screen. Don't stand there like a ninny! Can't you see you're letting in more flies?" Kate scolded each child.

Everyone seemed to find the doorway a good place to stand to recount the day's happenings at school. Grandpa too had little use for a shut door, nor did the flies seem to bother him. He always found, as soon as he'd opened the screen door, that something has been said which he hadn't heard.

"Whud'ee say, Kate? Whud'ee ay?" while he stood waiting to find out, door ajar, a thousand more flying pests swarmed into the warm, food-smelling kitchen to plague and annoy and fall in the milk, the syrup, or the butter.

In the late afternoon the car chugged into the yard, loaded to the running boards with men and their packs. They were as appraising of the family as the family was of them, but only one caught and held the children's attention and regard.

Billy Main was tall and rangy, and he walked with a free swinging stride as if the ground he trod on was his own. His hands were big and clean and kind, and his arms were strong and made for crawling into. His big grey eyes twinkled merrily and his wide mouth, full-lipped, was turned up at the corners. The moment the girls saw his crisp reddish hair and the way the freckles marched across his blunt nose, they all fell in love with him.

That first night his teacup was filled oftener and his wants attended to more solicitously than anyone at the table. Even Kate allowed him the biggest piece of plum pie. Sam liked him too, when he found out Billy was a good worker.

The other harvesters were a surly lot, but the children had little time to judge them in the course of the following events.

In the morning, Kate and the girls were busy filling up the gaps in the stockpiled baking on the pantry shelves. Grandpa was swishing pails of new potatoes around in a washtub with a broom and water. It made the peeling easier and quicker. A huge roast sizzled in the oven and filled the house with the odour of well-cooked meat.

It was time for the men's afternoon lunch. Kate and Grandpa had just packed a huge boiler with hot buttered biscuits dripping with jam. Sam came in to pick it up. His plans were as subject to change as the prairie wind.

"Don't worry about tomorrow, Kate," he announced. "They won't be here for breakfast. I'm going to pull onto Tom Dunn's quarter late this afternoon. It's right next to the north field where we are now and it won't mean moving the outfit so far. He asked me to thresh him just this afternoon." Sam poured out a mug of scalding tea from one of the grey-blue granite pots which had been filled and set to steep on the back of the stove. Grandpa, puffing at his corn-cob pipe, was emptying the tea into a five gallon cream can to be taken along with the men's lunch.

"He can't get anyone to do it for him, and I can't leave the poor beggar in the lurch."

"Whud'ee say, Sam? Whud'ee say?" Grandpa sat down a teapot and unclenched his pipe.

"Tom Dunn's," he yelled, " 'ee can't do that! 'Ee'll 'av trouble with the men!"

Kate stopped what she was doing, sat down on a chair, and groaned. The home threshing would be held up another week; a week if they were lucky and the weather held good. The crew would likely be disgruntled and hard to please when they got back. Worse still, there was the possibility that all Tom's cooking would fall to her.

" 'ur be queer in the 'ead. Men'll give 'ee trouble over it." Grandpa scowled, relit his pipe, and put it back in his mouth. He continued pouring and muttering dark predictions in fractured Cornish, through his clenched teeth.

Tom Dunn, was a "remittance man" from a wealthy family in Ireland. He had studied law, lost his first case, and after a mental breakdown, had been sent out to the prairies. He received a monthly stipend or "remittance" to insure his staying there, so as not to return and embarrass his family. A gentle, mousey man, he faded in and out of the prairie landscape, hardly displacing the air about him, and completely unnoticed in a group of tough western farmers. In this country where every man rode or drove a horse to town, no matter how close he lived, and where a "sheepskin" (fur-lined short coat) was the standard uniform, Tom was an anachronism.

In all seasons he walked across country five miles to get his mail or groceries. His slight figure, with meagre cotton smock billowing out behind him, could be seen breasting the bitter winds of winter or dropping occasionally like a slow motion movie, as he slipped on the sheets of ice that peppered his path on the flat plains to town. The only concession he made to cold was an oversize leather cap with earlugs which flapped down over his tiny ears, and a pair of huge woollen mitts with leather covers. When the weather was very cold he often stopped at the Marks' house to thaw out.

He was always extremely courtly to Kate, who gave him tea or hot soup at once. The older children were inclined to giggle at him and make fun of the icicles that always seemed to hang from the end of his sharp, patrician nose. The young ones loved and crawled over him, poking at his smock pockets for peanuts or candy hidden there for them.

"No more seats on the grandstand," he would say gently to the latest baby who tried in vain to crawl up on his already overcrowded knees.

Kate accepted Sam's decision about Tom's threshing and went on with her work, but Grandpa stubbornly kept objecting.

" 'Ow'll 'ee ever get the men into 'is 'ouse, Sam? Tell me that!" But Sam was gone out the door and back to the field as if he hadn't heard.

From the dozen or so heaps of rotting manure which dotted Tom's house yard, his living quarters differed very little. Only the rays of the westering sun that reflected on the sliver of glass in the tiny west window, and the worn wooden door that sagged on old leather hinges in front, designated his house from the other piles. Each fall he fortified the walls against the cold by pitching on another load of steaming manure. This assured him of a frost free winter.

That evening, following supper at home, Sam pulled the threshing outfit into Tom's field. With it went the bunkhouse on wheels which bedded down the extra threshers. The work-weary men washed their feet in turn in a large bucket in the yard, rinsed their socks in it, and then rolled their last smoke of the day. They fell onto their straw-ticked bunks as the harvest moon, orange as a Hallowe'en pumpkin, sailed up over the stooked stubblefields.

Before daylight Tom boiled coffee and fried platefuls of eggs which he put on his one-man table, along with loaves of bread and the butter, where the men would help themselves. Two or three nailkegs were the only chairs. The only available place to sit, besides the dirt floor, was the sagging wooden bunk where Tom slept.

The men, sleep and threshing dust still glueing their eyes, tried to squeeze into the one-roomed soddie. They were met mid-air by a covey of Tom's pet hens' The birds in alarm flapped in all directions, dropping feathers in the eggs and on the table as they flew about until Tom had shooed them all outside. Their usual place to roost, usurped by the men, was at the end of the bed.

"I train them to roost with their tails out," explained their master. "I always turn them to face me at night, if they haven't remembered — company, you know."

The hens were poor scholars, for their droppings were caked on the bedclothes, as well as the hardened ridge permanently plastered to the floor boards at the end of the bunk.

A disgruntled group of men made off to the field and were hard at work by sun-up. Ten o'clock tea-break came and went. Sandwiches or hot biscuits and tea were usually served to the ravenous men at that time. It was almost eleven, and no sign of Tom. The men were beginning to get angry as they drove in, unloaded their racks of sheaves into the threshing machine, and looked in vain for food. At last Tom's wispy figure appeared at the top of the rise, with the pail the men washed their feet in the previous night.

"I thought you might like some warmed spuds," quavered the frail little man, setting down the pail which was half-filled with boiled potatoes.

A big, dark, shambling fellow who had refused to go back to the field after half-past ten, jumped down from his rack and yelled to the knot of men gathered about the threshing machine, waiting to unload.

"Let's hang the stupid bastard!" He grabbed a length of rope from the front of his rack, seized the offending pail and was about to dump it upside down on Tom's innocent head, when Sam rushed forward and intervened. He planted himself in front of the bully and stared him down.

"We'll shut 'er down and go home. The Missus will have dinner for you," he promised, averting a riot.

CHAPTER 45

Saturday morning buzzed with activity. Jobs left until the children would be at home to help, had to be done. The imminence of threshers increased the tempo ten-fold.

Grandpa dragged the round wooden barrel churn from its corner to the centre of the crowded kitchen floor. The churn lid, encircled with iron, was held in place by iron arms which folded down and had to be carefully fastened when the barrel was full of cream. It smelled faintly sour the moment the lid was removed. The inside always had a damp, oily appearance no matter how carefully it had been washed and scalded after each churning. The handle at the side had to be turned at a steady pace so that the cream inside could slap and roll in order to coagulate the butter more quickly. It was boring and monotonous to the handle-turner.

From the cellar Grandpa lifted up a ten gallon crock filled with souring cream, and covered with a cheese-cloth lid. The moment the cheese-cloth was removed, swarms of flies lit on the edge and sides, or fell into the cream.

"Gor'dairney all!" He waved his arms in exasperation. "Stay out uv it! Stay out, I say," he blustered at the unhearing flies. He grabbed a rounded saucer-shaped tin, filled with holes, and began to skim the pests from the top. Kate helped him pour the cream into the barrel churn.

Mary did the breakfast dishes, and her high soprano voice sailed out on the crisp fall air. "I Love You Truly," she sang with more than normal gusto. Ruth helped stack away more baking and Hazel, supposedly practising her music lesson, drifted dreamily into "Melody of Love".

That's not your lesson, Miss!" Kate's ears, tuned to the multiplicity of activities, sensed deviation at once. "Get off that piano immediately. You can come and do this churning!"

She set Hazel to work.

"You hold the churn like this, and let the gas out by pulling the plug at the bottom here. You have to stop every few minutes to let it out, mind," she instructed, "and be sure to pound in the plug good, so it won't fly out and spill the cream. Be sure to put the lid back on carefully if you take it off."

She pulled out a big tin trunk, lined it with newspapers, and packed in the loaves of fresh baked bread.

"All this food! Drat the luck. We'll have to do it all over again — this will be all dried up by the time the men get back here again to thresh."

She wrapped the top with clean sheets. The dry prairie air quickly hardened eatables made with flour to rock-like consistency. Waxed paper was an unknown luxury. Grandpa peeled several potatoes and cut an apple in half. He put them in on top of the bread under the sheets before closing the lid.

"It will keep fresher. If it gets too hard we'll dip'um in water and cook over — makes like fresh loaves. Never mind, Kate, 'em'll be 'appy to get it dry or no dry, after Tom's."

The anxiety was wasted. The supply of food was considerably depleted by sundown.

"I just love curly hair and freckles on a man!" dreamed Hazel, slowing the handle of the churn and gazing out the window towards Tom Dunn's, where the object of her affections, Billy Main, would now be pitching sheaves and probably singing as he pitched.

"Hazel, will you keep that churn going! The butter will never come — and don't forget to let out the gas." Kate's hands were now busy with the younger children, as well as the immediate problem of what to do with too much prepared food.

The churn began to rotate slowly.

"I hope he comes here tonight. Maybe we can have a sing-song. I wonder what pieces he likes. Do you think he will go to church with us tomorrow?"

Mary stopped her singing and put away the last dish with a flourish.

"I'm sure he'll be here to go to church." She got out the cake ingredients and stoked the stove again.

"I'm going to make some chocolate cake. Don't any of you kids take any. It's for tomorrow!" The pantry shelves fairly groaned with baking, but there was no chocolate cake left. Billy had eaten three pieces the night before, showing his preference in cakes.

The flip-flops of the churn slowed and stopped.

"Keep that churn going!" Kate rushed through to scold, just as Sam, with the car filled with men, drove into the yard.

"What in heaven's name now?" She peered out the uncurtained pantry window. The happy rumble of Billy's laughter floated in. The magic sound galvanized Hazel into action. She whirled the churn, spinning the barrel. Sam and the men filed into the kitchen at that moment, and Tom's deficiencies were briefly explained.

The gas in the churn, neglected too long, took this moment to pop the cork. The cream shot out, struck the ceiling, walls, floor and splattered over the men trying to dodge it, before Hazel could slow down the rotating churn.

Grandpa rushed through the flying cream.

"Gor' dairney all! Can't 'ee watch wot 'ee's about?" he roared, waving his arms in the crowded melee. He grabbed a mop and flung it into the widening ring of what should have been the week's supply of butter.

For a moment Kate stood overwhelmed. Then she charged into action.

"For heaven sake, Sam, take the men outside until we clean up this mess." She whirled on the careless churner. "You get up those stairs, young lady," she ground out under her breath, "and get all the beds made by noon. I'll attend to you later!" Hazel sat, red with embarrassment, in front of her hero. Billy stood in the centre of the room among the men, plastered with cream, and laughed.

"I've had enough of this! I quit!" The big trouble maker scraped off his face, flipping the clots of cream on the floor. He stamped off to the bunkhouse to collect his pack. Several other men followed him and with two days' pay in their pockets, they made off across the fields towards town, leaving only half a crew for the afternoon threshing.

"Probably I.W.W.'s," Sam growled later, as he washed up for noon dinner. "Men got pretty mad at Tom's. Hated to refuse the poor beggar. Can't leave a man's crop to rot in the fields." He yanked down the long roller towel hanging on the kitchen wall. "Nothing much else I could do."

Kate sighed, accepted the change in plans, and swung into the dinner preparations. At least the food wouldn't be wasted now.

The weekend passed joyously for all. Sam collected enough new men to continue harvesting. Kate's food supply got eaten up, and Billy went to church with the older girls.

The younger children were just as fond of the handsome harvester as their older sisters. He brought a new song sheet with him called "Beautiful Mickey", the popular song of the times. Everyone learned to sing it around the piano after work. On the front of the music sheet was a picture of a girl riding a long-legged black horse. Every time Billy saw Edith riding Old Jim he called, "Hello, Mickey" after her. She clung to him like a wet leaf in the house, and tagged after him everywhere outside.

"You go for the cows," instructed Kate. It was Monday after school. Edith, impatient to get out to the threshing field to see Billy again, galloped Old Jim all the way to the pasture and rounded up the cows.

"Get going!" She reached down and switched the lead cow a swat on the rump as it ambled towards the ravine trail home. The cow waddled into a clumsy trot, ankle hooves clicking and milk-heavy udder swinging, followed by the others. Edith did her best to encourage a stampede until they reached the rise in front of the house. No cow was to run before milking. Riding with Billy and driving his team was too exciting an experience to worry about obeying a few rules about cows.

When the herd was safe in the barnyard, she closed the gate and sneaked down the outside cellar steps for a couple of apples out of the new barrel just brought home. She climbed on Old Jim's back again and galloped all the way to the threshing field.

Billy was eating hot biscuits and tea on his lunch break when she rode up.

"Hello, Mickey!" he called. Sitting on the side of his rack, some way off, with the reins of the team in one hand, Amy sat, sucking her thumb and happily swinging her bare legs. She had stayed home from school and had had Billy to herself all day.

Edith got off horseback and tied Old Jim to a tree, took out one of the apples and chewed off a bite, wondering how she could get rid of her sister. Amy took her thumb out of her mouth, stopped swinging her legs, and said, "C'n I have a bite of yer apple?"

"Oh, I dunno. There's a whole barrel full down the cellar at home." Edith crunched slowly, trying to make the bite last.

Back went the thumb in the mouth. Amy eyed her wisely.

Edith took the other apple out of her pocket, tossed it a bit, then took another bite from the one in hand. "I have this other one, though."

Out came the thumb. "C'n I have that one, then?"

Edith crunched some more. Billy must be almost finished his lunch by now.

"I'll give you my three gopher tails for it when I get home," Amy bargained, watching anxiously as the apple disappeared.

"Why don't you go and drive John's team for a change?" her sister bridled, stalling for time and trying to be casual.

"Nope! I like Billy best!" Back went the thumb in the mouth. Billy stood up by the lunch cans. The situation was desperate.

"If you go and drive John's team, I'll give you this whole apple to yourself."

Billy wiped his mouth, put on his leather gloves, and was striding across the stubble towards them.

"O.K." Amy reached for the apple and handed over the reins. John's rack was moving off towards a distant row of stooks, but she paid no attention to it. She bit into the apple and continued to sit on the side, still swinging her legs comfortably.

"Hurry up! There goes John now." Edith tried to shove her off the side.

"I ain't going!" Amy crunched away with unconcern and smiled at Billy as he leaped over the side and took the lines.

Edith was scarlet with rage.

"What's wrong, Mickey? What's the trouble?" queried the beloved. He peered down at her face as the horses trotted off across the soft black dust towards the open field.

Edith was speechless, but Amy wasn't.

"She wants to ride with you by herself," she grinned up at him. "She's mad cuz she gave me an apple to ride with John, but I do' wanna." She took a big bite of her apple. "I like you too!"

Billy threw back his head and roared. "Never mind, Amy. You can stay if you want to." He put his big arms around Edith and gave her a comforting hug, broke into glorious song, his singing almost compensating for not having him exclusively.

"Mickey! Beautiful Mickey! With your hair of golden hue! In your smiling, so beguiling! Can you blame anyone for falling in love with you?"

He sang all the rest of the time just to make up for it.

Everything seemed pretty right for the rest of the day, but something had to be done about that double-cross. Edith secretly resolved to pour water in her sister's dresser drawer to get even, as soon as they got home.

Billy loved them all, but he seemed to spend more time with Mary in the front parlour with the door shut. He took her to local dances, and his attentions to her didn't seem to impinge on the others' rights at all. Edith and Amy had the wedding all planned, and envisaged a kind of harem-like arrangement where Edith would be called "Mickey" and be sung to all day long, while Amy would have a place on Billy's knee and a constant supply of apples to eat.

Harvest time was soon over. Billy swung up onto the train one early winter's night, and headed east. The first week without him was bleak with monotony. It took a long time for his presence to fade from memory. At last only the happy sound of "Mickey" being played or sung brought back thoughts of the handsome young man who loved everybody and was thoroughly loved by all in return.

CHAPTER 46

"Good provider? Of course my husband is a good provider!" Kate snapped at a door-to-door washing machine salesman, one of the swarm that descended on the prairie farm wife after the war.

"He's got to be! How else would these children get fed and clothed? And he does a good job of it, too." She added this last threateningly. Following some formula of the day, the salesman had tried to slip in a note of sympathy in order that his intended victim might feel she was hard done by, and so buy his product.

It was the wrong technique to try on Kate. No technique was any good if it was buying something for the house she could possibly do without.

"I have the first mechanical Maytag washing machine that was sold in this country. It works fine and my husband keeps it in good repair, thank you. My husband likes mechanical things, and if he can't fix it, it can't be fixed!" She flounced back to her work, leaving the salesman to find his own way out.

The tiny engine which powered the Maytag had long since worn out. With his usual inventiveness, Sam ripped it out. He trundled up another engine, almost as big as the washing machine itself. With the men's help it was turned on its side and manipulated through the kitchen door. He connected it to the washer by a big belt. A few spikes pounded into the linoleum held it in place, and with a twist of the crank it roared into life. It shook the house to the rafters with the noise and vibration, and filled the kitchen with thick blue smoke.

"Works like a charm, Missus!" Sam stood over his invention with all the joy of the discoverer of the wheel.

"More horsepower than the other one. Should do the washing twice as fast," he yelled.

The machine did the job all right, with reservations as to its' efficiency. Any discrepancies Sam didn't notice, and Kate was too busy to point out. Besides that, she had complete faith in her husband's ability.

The engine took up most of the room not occupied by the washing machine, so that no activity except the washing was possible if the engine was going. Getting the clothes in and out of the tub required some expertise too. The wide spinning wheel could catch and tear your skirt off if you weren't careful.

Another disadvantage was timing. The engine was used for the chopper, the machine that ground up grain for the pigs. If the men decided they needed more chop, they unhitched the belt and moved the engine back to the barn, no matter what stage the wash was in. Tubs of congealed water filled with half-cleaned sheets or Sunday clothes weren't in a class with a barn full of squealing pigs.

Other discomforts were the gas fumes that belched out of the engine, filling the kitchen with blue smoke, well laced with carbon monoxide. In the winter time these fumes fanned into the rest of the house and caused everyone to have a headache on washday. Fortunately, with the constant traffic through the kitchen letting in bursts of bracing icy air, everyone was saved from asphyxiation.

One day Kate was sick, so Sam did the washing for her. The next day he set right to work to improve things. He sawed a three-foot square hole in the kitchen wall. The men put the engine outside, and Sam threaded the elongated belt through the hole and attached it to the washing machine inside.

"Works like a charm, Kate," he hollered, "come and see! Lots of room now, and not a bit of gas in the kitchen! Just take out the belt and put this canvas over the hole when you're not washing!"

He was right. The nauseous gas was whipped away on the winter wind outside. But now there were other elements almost as dangerous inside. Every drop of water that sloshed over onto the kitchen floor froze immediately into sheets of ice, extremely hazardous to the unwary washer.

Spring came soon, and the ice on the kitchen floor melted. The whole experiment looked such a winner that Sam decided to rig up the churn, too, with a labour-saving device involving the engine.

"I don't know, Sam! It has to be watched all the time. I'd rather the girls did it. We aren't getting much cream these days, and I'd hate to lose this lot."

"Don't worry, Kate. Father will keep an eye on it, and it will save all that quarrelling over the job!"

In a few hours the churn was attached to the engine and ready to make butter.

"Bring in the sour cream, Father, and we'll see how it works."

The cream was tipped in and the experiment began. Mary was set to watch it, with Grandpa overseeing the job. The churning progressed with success. Sam, pleased with himself, rushed off to the barn, Grandpa turned to other jobs and Mary, impatient to be free, took off the belt and the lid, to see if the butter had come. Kate, busy in the kitchen, shouted a warning.

"Watch how you put that lid back on, Miss. Be sure it's back in the right place!"

Her warning came too late. Mary attached the belt, the churn made one revolution. With a clatter like thunder, away went the lid, the cream, and Sam's mechanized churn experiment. Mary escaped upstairs until supper was over.

"If Sam could put an engine on Old Jim, I'm sure he'd try it if he thought the horse would trot faster," Kate complained to her father-in-law as she mopped up a week's supply of butter, and surveyed the waste of her careful saving.

She had been brought up in a Scottish porridge-and-potatoes household, and wouldn't spend money for anything if she could help it. Sam's concept of living was to think big and act fast, which resulted in lots of land, lots of machinery, lots of kids, and lots of mistakes.

His record of good provider had many vicissitudes — some good, some bad. A few of the best were a Weber piano on which all the children learned to play; a set of heavy oak furniture which had a huge buffet, a table that seated twenty with ease, twelve chairs, and a heavy lumber bench along one side which could hardly be moved to sweep; and a bed-lounge which withstood the determined daily attacks of the destructive family for twenty years.

When Sam ran into things he thought Kate would need, he bought them on his own, sometimes with dire results.

"Don't put any boots on the Eaton's order for the kids, Missus," he said to his wife when he got home from a trip to Winnipeg one fall. "I've got a whole packing case coming! Got them real cheap, too!"

"But, Sam, what about sizes. They mightn't fit!"

"Don't worry about that, Kate. I bought lots of them. If one pair won't fit, they'll have a choice, and they'll be warm — all fleece-lined, too!"

They all waited in some anxiety for the parcel, especially the girls. There were no fleece-lined shoes for girls in the catalogue or in any other store, then.

When the boots came there were lots of them, and they were all fleece-lined, but nobody wore them.

Kate opened the packing case, and it contained fifty pairs of ravelled and shrunken boots, with wide front lacings and flat, hard leather soles. They were fitted with cleats for adjustable skates, which had disappeared from use before any of the family were born.

"Now we'll say nothing about these," warned Kate. "Some of those crooks in the big stores sold them to him from a fire sale, and all Dad could see was that they'd be warm."

During the flu epidemic Sam saw more bedding was needed. When the epidemic was past he went to Brandon with a load of pigs, and did some shopping.

"I found an Army Surplus Store, Missus," he announced jubilantly when he got home. "It's stuff made for the soldiers' use and left over from the war. I got blankets — all we'll need for some time — and cheap, too!"

The children waited breathlessly for their arrival, and when they had put the new blankets on the beds and got under them, they wished they were breathless. The blankets smelled as if they had been manufactured from recycled tar-paper, with generous shots of diesel fuel. They were heavy, stiff, an ugly brown-black colour, and had a built-in refrigerating system.

"They'd freeze the nuts off a gang-plow," remarked a hired man, after the first night spent under their crushing weight.

They experimented several ways to make them work. On cold nights the children doubled up. Four kids in one bed with two bedlots of blankets were a pretty sure thing they'd be warm by morning. There were disadvantages to this system. The little ones who shared the beds were not night trained. In the morning, from sheer weight of the blankets above, they were warm, but if someone made a move, the winter air would creep in and turn the warm pools of dampness icy cold around them.

They tried another experiment. Trixie, one of the fox terriers, was coaxed into bed with them, but they only were cosy for a few nights.

"Mama! There's bugs in these blankets! And Trixie's got bugs, too. What'll we do, Mama? We can't sleep!"

The "Battle of the Bugs" was a constant in Kate's life. The bed kind, red and smelly, that got into mattresses and walls and attacked at night; and lice, or "greybacks", that got into heads and clothing, were a never-ending problem. Neighbouring bachelors, harvesters, and the children of harried mothers like Kate, all left their own particular brand. The cross-breeding of these developed a tough strain which seemed to thrive on heat, cold and poison.

Every Saturday (bath) night heads were doused with coal oil. Every tub of washing was well laced with creolin, an antiseptic for horses. Kate froze the underwear on the line after boiling it. This process killed the walkers, but the nits hatched and in a week the greybacks were back in business, hungrier than their boiled and frozen grandparents for a chew at children's itchy hides. Occasionally Kate left the house closed for a day and fumigated with formaldehyde or burning sulphur, both cures bound to kill the toughest bug, and bound to make anyone else think they were dying, too, if they got a whiff of the fumes.

Kate made a quick check of the new blankets. It was clear the bugs liked their acrid smell better than she did. The blankets were loaded.

"Get the fire going under that vat in the yard, and fill the top with water — even if you have to bring up two barrels full!" she directed the boys one windy spring day. "I'll boil the blankets with creolin and they'll dry in an hour in this wind!"

Time went by, and Kate tried to immerse the first blanket in the boiling cauldron. It was so bulky and so hard, it wouldn't go down in the water. After a struggle, she came back to the house with purpose in her step and fire in her eye.

"Gather up all those new blankets — every one one of them, and bring them out to the vat," she ordered.

One after another she jammed them into the firebox under the vat and they exploded in flames like a molotov cocktail.

"Now you girls get out all those woolen things we've been saving for quilts. We'll just get to work on them right now."

When Sam came in for dinner she announced, "Sam, I've burned all those accursed blankets. It's a miracle the Germans didn't win the war, if that's the shoddy stuff our soldiers had to sleep under!"

CHAPTER 47

Kate's pregnancies gave her very little time to do any community work, but what she did counted. Her most exciting effort was as a delegate to a "Farm Women's convention in Saskatoon. More localized were her buggy trips to various neighbour women's homes to help at a birth. After her first experience as a midwife at the Larches', pregnant or not herself, she answered every summons.

First came a brief phone call. "Get out the horse and hitch it to the buggy," she would command. From the clean clothes basket, she rescued un-ironed sheets and clean rags, packed them in the ubiquitous flour sack, and was off by herself, holding the reins stiff-armed out in front of her, and clucking nervously to the somnolent horse. She never returned from these errands of mercy for at least two days, and the family were to "carry on" efficiently in her absence.

She was the "Carrie Nation" of southern Saskatchewan before prohibition came into force.

"No one the worse for liquor need darken our door." This oft repeated warning was well understood in the neighbourhood, and few tipplers ever dared to challenge it. She never missed a meeting where the evils of drink were discussed, if she could help it.

The day they voted on Prohibition, Kate hurried through the morning tasks so she could go to vote in the afternoon.

"Can't go," blustered her husband, "have to get that engine fixed — went on the hummer this morning. Got to keep the men busy. Father will take you," and he rushed off to the barn.

"I'll 'av to 'elp 'ee, Sam," his father shouted after him at once. Grandpa Marks' always anxious to break from farm monotony by a trip to town on any pretext, was strangely reticent about going today, but he hitched up the horse for Kate.

"Hmmmnph," sniffed Kate suspiciously. She took the reins and was just about to drive off when her Cousin Robert appeared at the gate.

"Prohibition voting today? Ay! Ay! Katie Cameron, I'll go with you." The old man struggled up over the wheel, and they drove off together.

"Well," she exulted, as she climbed into the buggy and helped her cousin in for the trip home, "there's two good votes against the demon rum, anyway!"

"Och no, Katie Cameron, only one!" boomed the brave little man. "Only one!"

Kate gasped. "Cousin Robert!" Do you mean to tell me you, a man of the cloth, YOU voted AGAINST prohibition?"

"That I did, lass. That I did!" he replied calmly.

"But why? You know what an evil this liquor is."

"That I do, lass. That I do. But in my considered opinion it's a man's inalienable right to go to hell in whatsoever way he wishes. I voted against prohibition."

Kate was quiet all the way home.

For almost twenty years the cradle in the house had never been empty. No sooner did one crawling infant stagger to its feet and begin to toddle, than another one arrived to be diapered, burped and baby-bathed.

One blustery April afternoon mysterious sounds, hurrying feet, and snatches of hurried conversation floated down the stairs from above where Kate, a doctor, and a nurse had spent most of the day.

"I'll bet you didn't expect to have all these children when you came west, did you?" The doctor busied himself with the activities of child-birth.

"A baker's dozen, isn't it, Mrs. Marks?" and Kate's thirteenth child was born. "This one should be called Lucky," he said, as the father of the child came into the bedroom.

"I'd like to call him Robert, then, after Cousin Robert. It was through him I came west in the first place."

Kate, weak but firm, looked to Sam, for affirmation. He had just watched, white and shaken, while the process of bringing life to his newborn son had taken place.

With a sputter and a snort reminiscent of the Reverend cousin's pulpit oratory, the baby let loose with a loud, bellowing howl.

"Couldn't be better named!" gasped Sam, with a relieved grin. "Couldn't be better named!"

Kate's last child, Garnet, was born two years later. His arrival was so sudden that Sam was alone with his wife when the baby arrived.

"Maybe that's why we didn't have any more children," Kate teased him later. "Now you know how they are born!"

CHAPTER 48

The trauma of death in other homes as well as their own was shared, mourned over, and part of the family's upbringing. Funerals weren't oak boxes, flowers and high tea after.

One cold March day a cutting wind had driven sleet against the windows all morning. By noon the wind died, but the thermometer stayed well below zero.

"The Smith's girl's funeral is this afternoon." Kate hustled back and forth, putting loaves of bread in pans and giving orders as she went.

"Take the bread out of the oven when it's done, girls. And don't let the fire die down. Don't put in too much wood, either. You boys get a barrel of water up on the stoneboat, and you can pile up that wood if you've any time left over. No quarrelling, mind. We'll be away all afternoon, like as not."

She rushed off to change her clothes, followed by Margaret and Violet.

"Can't we go to the funeral, Mama? You let Kitt and Pat go last year! Can't we go, Mama? Please?"

Kate paused. "Hmmm. Yes, you may go."

Don't you think they're too young? Kind of a shock. . . ." Sam was putting on his best suit in the bedroom.

"It won't hurt them. Death is a fact of living. It's a natural thing."

The two little girls, dressed in their Sunday School clothes, went along with their parents to the neighbours' home, where the funeral was to be held. They sat quietly in the solemn gathering and stared with wonder at the tiny, waxen face of their recent playmate in her pretty white casket.

After the preacher's service, the church organist took her place at the old family organ. "Jesus Bids Us Shine," the child's favourite hymn, and "Abide With Me" for the comfort of the parents, wheezed out. Muffled fitful singing kept a sad accompaniment.

After a prayer the crowded room emptied. The silent neighbours followed the casket, borne by four young boys, to the sleigh where it was placed in a plain pine box. Families climbed into their respective sleighs and cutters and drove off in a slow moving cortege to the windswept graveyard five miles away. The horses were tied along the fence, and the people made paths through the snow toward the mound of clay that scarred the white expanse of the burial place. Fresh snow had already drifted into the gaping black hole in the ground.

The biting wind congealed tears on the faces of the mourners. The minister finished his address and intoned, "Ashes to ashes and dust to dust." The little box was slowly lowered into the earth. Clumps of frozen sod thumped down eerily as the men began to shovel in the dirt. Others silently walked back through the snow toward their horses and rigs, and the funeral was over.

The two girls climbed into the sleighbox and cuddled down beside their mother on the hay in the bottom of the rig.

"Don't cry any more now." Kate covered them with blankets, wiped tears and noses and gave each a firm hug.

"She'll be so cold down there all by herself."

"There now. She'll be all right. It's God's will." The muffled chink and jingle of the harness were comforting sounds as the team plowed into the darkening drifts towards home.

Kate held warmly to her two children. Her thoughts wandered from the grief her neighbours must feel over the loss of their child to a private anxiety of her own.

Her Cousin Robert's declining health was a nagging worry. His education, knowledge and Godly example had been an inspiration to her in the rearing of her children. The long walks the old minister took across the plains he had loved and served so well for half a century had become less frequent. He now gratefully accepted rides wherever he went, and his visits to the family were rare.

"I know I should offer him a home, but where to put him in our crowded household? And the children — would their noise bother him? And Sam doesn't understand him," she fretted.

Several days later news of the death of the Widow McLean reached Kate. The old log house which had sheltered her ancient cousin was to be torn down. To her great grief she heard that he had immediately registered himself in an Old Men's Home some fifty miles away, and moved out of the community.

One evening months later at the neighbours, the McRae's, a loaded wagon drew into the yard and disgorged the old gentleman, along with some boxes and an old wooden trunk. His hearty, "God bless you!" had lost none of its ringing sincerity, but the old minister's clothing hung listlessly from his shoulders, and his skin looked yellowed and drawn. He tottered into the familiar kitchen, surrounded by the warmth of long-loved friends.

"Welcome back, Mr. McDougall! Welcome back! And you'll be staying with us the while —," and members of the family carted boxes, bundles and the trunk into the back "company" bedroom.

A few days later the old man became very ill. He passed into a coma, rousing from time to time to mutter prayers in Gaelic or English. Days passed, and it was evident the end was near for the tired old preacher. He was quiet for hours, except for laboured breathing.

Suddenly he began to speak in a youthful voice, repeating words of love in lilting, liquid Gaelic.

Then, "Lucinda!" he moaned. "Where is she now? Surrounded by wealth and comfort, her children's children at her side — and where is old Robert? Ah, roaming the prairie and surveying the air —"

As the warmth of life began to leave his body, the nurse placed hot water bottles in the bed beside him. He stirred again, and plucked at the bedclothes.

"Warm — warm? A woman? —" he asked wonderingly, "a woman —" and the old orator, who had foresworn an earthly love for the life of a celibate in order to serve his God with greater dedication, fell asleep, never knowing.

Kate accepted the news of his death calmly.

"He's lived a good, useful life," she remarked, dry-eyed, as she prepared to go to McRae's to pay her respects. "He'll be remembered long after the rest of us — as an historian. He's written about this area since the eighties."

By the time she arrived, her Cousin Robert had been bathed and dressed in his clerical suit by the loving hands of his old neighbours, and "laid-out" on boards on top of the bed in which he had died.

"His diaries —," faltered Kate "he promised them to me long ago—" She joined the family and friends in the front room.

"Not much left," supplied the nurse. "He burned many old papers the night before he took sick." She gave Kate all that was left — a school exercise book, with only a few entries in it, covering his stay at the Old People's Home. Some comments concerned the death of inmates. Others were about the difficulties he was having with his worn-out body.

May 8th — John Quin — promoted to Glory — prayed for him.

May 12th — Mark Abel — called Home this morning — none of his family present.

June 15 — No B.M. since Sunday week.

June 20th — Wonderful deliverance between lunch and tea.

These mundane bits, written in his precise, spidery hand, were all that remained. Only one page, closely written, showed a spark of his former depth of spirit, and appeared at the end of a commentary on a meeting of world leaders.

"They have made great progress in the manufacture of the implements of death. Could they not do just as well in the pursuit of World Peace?"

Kate put down the book, took out her handkerchief, and wept.

"I suppose with age our machinery wears out," she said, with resignation, "and we're all reduced to concern for our bodily functions. But his writings! — what a loss!"

"Perhaps he destroyed them for fear something he wrote might hurt someone — but he left wonderful memories and an example —" comforted a friend. "Maybe other earlier diaries will turn up —" and several did, later on.

Kate went home, and was greeted at the door by three-year-old Robert. He put his chubby arms around her neck.

Don't cry, Mama. I'm going to be a preacher like Cousin Robert when I grow up."

It was prophetic comfort. Robert entered the Salvation Army Training College fifteen years later, and faithfully followed the example set by his namesake.

"Learning to accept death is as hard as learning to live right," Kate told her children. "we all have to face it sometimes."

Deaths that happened in other places were wondered at and forgotten. When they took place in the family circle it was harder.

One cold November morning Grandpa Marks, who had been ill for several weeks, passed into a comma. He awoke from time to time, muttered about wills, laughed a little, and said, "Tell Edie she may 'av the 'orse." He closed his eyes and his spirit winged away to the sound of the stormy winds he loved so well. His body, dressed for burial by a neighbour, was laid out in the icy parlour with the snow blowing in on him from the open windows. Three days later, friends of long standing gathered in the front room to sing hymns and listen to the preacher intone his requiem. He was trundled off in a sleigh box, and lowered into the frozen earth to lie beside his first wife in a snow covered patch in the Kennedy graveyard, with other pioneers far from their place of birth.

CHAPTER 49

Sam and Kate's dominant concerns for their family were first food, then shelter. Clothing came off a poor third. In a family as large as the

Marks, false pride was none-existent. If one could find a pair of holeless stockings, a coat with no buttons missing, and a pair of shoes which miraculously fit and didn't pinch, who would be silly enough to question their origin. Second-hand clothing was an interesting privilege, not a disgrace.

The first windfall was "Grandma's things", both the real Grandma Marks' and her fashionable successor's who left her clothes behind when she went back to England.

"These will do just fine for Edie and Amy." Kate hauled out two ladies' short fur coats, with stand-up notched collars, bulging busts and nipped-in waists. They had short peplums all around, which had looked very stylish over bustles in 1910.

The coats were sent away to be "made down" for the girls. The dressmaker took a stab at sizes, and the results weren't flattering. Sleeves had been shortened and shoulders made smaller, but there the "make down" ended.

Amy put hers on. The former waistline came to her knees, with the peplum fanning out below, giving her the appearance of a pouter-pigeoned dwarf with no legs. Edith put hers on.

Amy eyed her for a moment. "You look just like a toad," was her succinct description. They both developed a hatred for those cosy winter coats, and longed for a cloth coat from Eaton's catalogue like everyone else at school. It wasn't surprising that within a year the coats were conveniently lost and they got their wish — cloth coats like everyone else. They were cold, but they fitted in the right places.

Kate's courage in the face of all her hard work was much admired by her friends who had only one or two youngsters. They showed their concern by sending boxes of used clothing, and apologizing.

"I'm afraid my Mary is bigger than your girls, but I thought some of these things might come in handy —," or "Johnnie's grown out of these sweaters and overalls, but maybe they're too small for your Pat —?" they regretted.

"Thanks. The things will come in handy," Kate would reply. "If they don't fit now somebody's bound to grow into them in a day or two."

A beautiful young woman, only child of doting parents who showered her with expensive clothes, had married a man in a nearby district. She became one of Kate's admirers and best friends. One winter's day she came visiting, and brought a trunkful of treasures.

"Your girls will be going to dances soon, Mrs. Marks," she puffed. "Perhaps they can do something with these things. I'm too fat to wear them any more!"

Childbearing and the rigours of farm life had caused her to lose her lovely figure, and increased her weight. Her loss was the Marks girls gain. The trunk was packed with dresses; shiny pink satins, pretty pastel crepe-de-chines, lacy silks, and other things trimmed with baubles and beads, ostrich feathers and slinky furs, treasures as good as any Aladdin conjured up with his lamp.

Magazines and style books were dragged out, the sewing machine whirred constantly, and the girls set to work on the dresses to make them fit and bring them up to date. That winter Mary went dancing in a pink satin gown with an overskirt edged with bugle beads which swung and swayed at her every step. Ruth wore a silk pastel with maribou trim at the neck. At every party they blossomed out in new exotic creations from the magic trunk.

Twice a year the Eaton mail-order parcel arrived. In the fall, a huge packing case held doubles for everyone — long prickly stockings, fleece-lined underwear, and red felt boots. Fancy stuff like sweaters, coats and dresses were usually made-ups, made-overs and hand-me-downs.

The spring parcel contained clothing specifically for wearing to church — hats, white rubber-soled shoes, stockings, and yards and yards of sturdy, no-nonsense material like checked gingham. Dresses were cut out and sewn, with not much attention to the fine points of tailoring, but looking splendid in the final glory of the finished product. Nobody noticed a shifty seam or two. The boys were always lucky. Nobody quite got to know how to make a man's shirt collar, or tailor a man's suit, so their things were always "store-boughten".

Next to church, prairie picnics were important for showing one's best side. They were held after seeding and before haying, a comunity visit-in for farmers' families, where you could watch a ball game and discuss crops if you were grown up, or scrounge money from a parent's supply of nickels, if you were a kid. The picnics were held in wooded glens, on edges of towns, or at Vandura Lake. The lake was the best, because you could go swimming as well as make yourself sick on ice cream.

"Mary," Kate said one day, "the picnic's this week. Here's some material. See if you can make something for the girls. The rest of you gather up your shoes and stockings and see about your underwear, so you'll be ready when the time comes." She dug out a bolt or two of gingham. Mary grabbed a pair of scissors and the Ladies' Home Journal, and waded in.

"Edie and Amy, this is a good time for you to learn to sew." She threw them some old blouses trimmed with lots of lace.

"You get to work and pick off this lace, and you can trim your own underwear with it."

She dug in a basket, and pulled out several empty flour bags.

"Ruth, you hang these out on the line, inside out, mind, so most of the flour will blow out. I'll wash them later and cut out some pants and petticoats for the girls. They can sew them by hand."

"Mama! Why don't you buy Five Roses flour, like the Potters do?" Edith begged. "I just hate that Robin Hood kind."

"Drat that child!" Kate stopped in mid-kitchen. "Where ever do you get your ideas?" She shook her head and rushed on to the pantry. "I've more to worry over this morning than the imagined quality of the flour used in this household."

She had many surprises from her children, most of which she understood, but occasionally one stumped her. She came back and looked at Edith, puzzled. Had one developed gourmet qualities so young?

"Why don't you like Robin Hood flour?"

"It's not exactly the flour, Mama. Frances Potter always has roses on her pants. I just hate Robin Hood on mine!"

Kate's face assumed a judicial expression, and she turned quickly away.

"I'll see they're put through lye this time," she mumbled.

Robin Hood's male proximity to one's backside, and the direction he pointed his bow and arrow were embarrassing. It also called for rude remarks from classmates, all boys, which hampered freedom of movement and running ability at school on a windy day.

The dress material ran out, and Mary improvised. She took her last winter's satin gown, a few frills from an old crepe-de-chine blouse, a wide silk ribbon for the sash, and made Edith a dream dress. The day of the picnic Edith put it on and jumped in the back of the McLaughlin with the rest of the family.

Kate pinned a big wide white lace hat, splendid with yards of ribbon and feathers and flowers, on her piled-up hair. She had sent to Eaton's for a large-size hat, on approval. It had arrived, and cost the astronomical sum of twenty-two dollars.

"I'm keeping it!" she snapped at her husband, when he suggested maybe she should buy a dress instead.

"None of my dresses ever fit for long, anyway. At least my head size doesn't change!" and she stubbornly hung onto it, doing without other garments so she could for once have something of sublime beauty and utter uselessness.

She got into a blue silk dress, belonging once to the fashionable Grandma, which miraculously fitted her (she wasn't pregnant that summer), and climbed into the front seat with the last baby on her knee.

Sam up front, with his pockets full of nickels he had got at the bank the day before, wore his best black suit and homburg hat. He honked the horn incessantly.

"Anyone not ready will have to walk. I'm leaving right now!"

After a last nose count, away they went, down the road and over the hills, with hearts light and colours as bright as the flowers blowing on Kate's hat.

Sam drove into the picnic area and lined his car up with others around the ball diamond, where a rousing game was taking place between two neighbouring districts.

The children opened the doors and poured out of the back like colourful animals out of the ark. They skittered off to find friends and run foot races.

"Why, Mrs. Marks," exclaimed Mrs. Cairns, the wife of a farmer minister, and one of Kate's telephone friends, "your children are a sight to behold." Her round little face, wreathed in delighted smiles, popped up from the next car. "Solomon in all his glory was not arrayed like one of these!"

Edith found May, a pretty girl from the next district, sidled over to her, and waited for her to decide what they should do.

"Come over and meet Velma. She's my classmate at school," May invited.

Velma swung on the running board of her family's Ford car, and eyed Edith suspiciously as they approached.

"This is Edith. Would you like to come with us for ice cream?"

Velma made no reply for some moments, then said,

"My mother says it's not appropriate to wear silks and satins to a picnic!" she snipped. "Gingham is much more practical." She smoothed her gingham shift, looked disdainful, and waited.

May's pretty round face turned scarlet, and she turned towards her new friend. "Oh, Edith, I think your dress is just beautiful," she blurted, her voice full of an emotion that wasn't understood for a moment. Tears of hurt gathered, but were never shed. Edith rallied her forces and took the first step into an adult female world.

"May," she remarked innocently, "your car is a McLaughlin like ours. How do you like it? If there's anything I can't stand, it's a Ford. They are so noisy and so — cheap!" and she ran off to run and win a foot race.

At ten o'clock that night Kate was still holding a basin under Edith's head, and she was still throwing up.

"Sam, I don't know what you were thinking about, giving this girl all that money. You know what a pig she always makes of herself."

"What money?" Sam leaned over and looked down at the flushed face of his daughter in a puzzled way. "I gave them all the same amount."

"Same amount, nothing! I know her. She'd make three trips to everybody else's one. And how would you know which one was asking when you were talking?"

The sick child lay back and reviewed the day.

The pretty dress must have been prettier than expected, to judge from

Velma's ugly remarks. Then there were all the races won, to say nothing of the nickels acquired from her father.

"Young lady! Let this be a lesson to you!" Kate said, as she wiped off the fevered face and put a clean basin by the bed just in time.

"It's sure been a wonderful day!" Edith mumbled to her mother, in the middle of the next heave.

CHAPTER 50

Christmas 1922 was at hand. Kate managed to get the Eaton order away in time. On the happy morning each child received one gift and one larger gift was to be shared by everyone. Edith and Amy, too big for dolls, received books; Anderson's and Grimm's Fairy Tales. And grim most of those yarns were. Wicked step-mothers, werewolves, and the walking dead filled their winter days with fascinating reading, and their winter nights with dread.

"The Snow Queen's the scariest story," Edith said, after reading it aloud. "I hate the way she put an icicle into the boy's heart and then dragged him off on his sled to that terrible land."

The house present was a sled, too — a big red and yellow beauty with stout rope attached. Kate allowed the girls to take it to school one February day. She put it in the bottom of the sleighbox with the lunches and robes, to be taken out when they got to school.

"You two take care of it, mind," she warned. "Remember it belongs to everyone. If you leave it in town you won't get another turn."

That day lessons dragged for the two girls. Both were impatient to join the town children at hitching rides on the back of farmer's sleighs after school, a sport they had watched enviously for weeks. Permission from the farmer was first asked. Then the sled rope was slipped through a stet on the sleigh runners, the horses pulled out, and one flopped belly-down on one's sled, gripped the rope, and hung on. To keep from slewing into the roadside drifts, or from turning upside down in the ruts made by the runners or by the horses' hooves, was a skill to be learned only with painful but exciting practice.

If the team was fast, frozen clods hurled back from the horses' flying feet. The exhilaration of zipping along with one's face so close to the swift-moving snow beneath was worth the hits in the eye or the spills suffered along the way.

A warm chinook had blown all day. Murmuring water under the snow sang falsely of spring. The greyed-green sky clamped down on the world like an inverted cup. The wind died and an ominous air hung over the town.

The girls had fifteen minutes before the team was ready to head for home; time enough to get their sled and hurry over to the store with their classmates Geordie and Wilfred, and to hitch a ride or two on outgoing sleighs. Two teams with sleighboxes were tied to the front of Lowe's General Store.

"There's Chevalon's rig. We're gonna ask him fer a ride." The boys tied their sleds to the runners and made purposefully towards the store.

"Which one is Chevalon?" Amy whispered as they sidled into the dark interior.

The Chevalon name had acquired a curious fascination for the two girls since the fall before, after Sam and Kate had returned from a sale some miles to the south.

"I met the Chevalons today," Kate had announced when she got home. "Nice French people from Dumas. Just the one daughter, poor souls. Such a shame. She's sixteen, a grown woman, and still in her bare feet in public. Babbles and plays with dolls yet. Poor Mrs. Chevalon has to lead her around like a child."

For days the two girls had plagued Kate to visit the Chevalons. They discussed endlessly what this child-woman would be like, and burned with curiosity to see her themselves.

"For goodness sakes! Why are you both so curious about that Chevalon girl? She's a poor simple thing, and there's plenty like her, the dear knows!"

In town this day was the father of this changeling they had yearned to see. They followed the boys closely and peered into the dim recesses of the store where two men were dealing with the storekeeper.

"Chevalon's the little one," whispered Geordie. "Victor Mason's the big guy."

Chevalon, a wee, brown, nuggety man with a perpetual grin and the movements of a jumping-jack, hopped back and forth from store to sleighbox with sacks and boxes of food and fodder.

"Oui! Oui! One hunnert poun'. I take it easy!" He chopped out little phrases in French and behaved as if Mr. Lowe, the storekeeper, was a dear and personal friend.

A black leather cap, too big for his tiny head, came down over his eyebrows and ears. A long, black, dog-fur coat was tied round his middle with strands of binder twine, and hitched up under his chin, despite the mildness of the day.

"Merci! Merci! Sacre! Good flour!" and he heaved another sack, almost the size of himself, over the side of the sleighbox.

"Mr. Chevalon, kin we hitch a ride behind your sleigh?" Wilfred stepped in front of the little man. Chevalon threw back his head and roared as if the request was the wittiest of remarks and he was more than delighted to hear it.

"Oui! Oui!" he shouted, and slapping his sides he hopped into his rig and picked up the lines.

The other farmer, Victor Mason, a dark giant of a man, shambled out of the store and jumped into his sleighbox in front. He unwound the lines and crooked an arm towards the girls.

"You coom ride wit me!" he snarled. His greasy cap had a broken peak which hung over his wide, flat nose. Glittering eyes like black beads strung too close together shone out from under the peak. His untrimmed hair fringed over the collar of his mackinaw coat like wedges of old matted hay. He leered at the girls, his lipless slit of a mouth split in the semblance of a smile that showed stumps of discoloured teeth. Two tusk-like molars at the sides gave him the appearance of a predatory shark.

"Hurry! Hurry, or I go wit' out yuh!" he commanded. The two girls backed up and ignored the big man's threats.

Edith pushed Amy behind her. "We're going to wait for our team," she shouted to the two boys as Chevalon's rig pulled out. The girls dropped their sled rope and ran back to the safety of the store, the ugly man's voice still shouting threats in French after them. When the sounds of the jingling harness had faded they went outside again, to see Wilfred and Geordie tugging on their sled ropes, on the way back from their ride.

"Whyn't you come?"

"We didn't want to ride," Edith lied, ashamed of the wave of nausea and terror her imagination had created at the sight of these two strange men. Something about them had brought to her mind the story of the boy frozen to his sled by the wicked Snow Queen in her Christmas story book. She wondered if it was the menace in the big man's voice, or the sight of the two boys going off on their sleds that made her heart thump against her ribs in fear and panic. The home team coming around the corner was never so welcome.

"I'm sure glad we didn't go with that ugly man. He kind of scared me." Amy clung to her sister's hand and they ran together to get into their own rig.

On the way home the temperature plummeted, a common prairie occurence. Puddles of water hardened to ice. Winds gusted across the road and knifed through coats and mitts. The children huddled together in the bottom of the sleighbox for warmth. They arrived home shaking and stiff from cold.

By next morning the worst blizzard of the winter swept across the plains, wrapping the house in a shroud of stinging ice pellets. Roads were drifted in and impassable. Visibility was limited to a few feet.

Sam stamped in from the barn, shook the snow from his coat and set down two pails of milk, already rimmed with ice around the top.

"Pump's frozen. Over forty below with this wind, too. Wouldn't send a horse out in this storm. No school for the kids today."

All morning the raging storm blanketed the windows and made mournful sounds on the telephone lines attached to the corner of the house. By the afternoon the children had settled down, playing cards, thumping on the piano or quarrelling among themselves until bedtime. The telephone lines were down and they were cut off from the rest of the world.

The third night of the storm the wind lessened and the stars began to come out.

"Up to bed with all of you, now. The weather's bound to clear up by morning and it's off to school with all of you so I can hear myself think. Dad will keep the fires going again tonight so we won't all freeze." Kate scattered the little ones up to bed, lit another lamp and sat down by the table to read and keep Sam company. He sat in a chair nearby, reading the paper and putting blocks of wood on the fire. From time to time he fell asleep and snored.

The children undressed in a minimum of time and jumped into their beds. They snuggled down, shivering from cold and half frightened by the flickering patterns the scudding clouds made across the moon in the black night sky outside.

Suddenly, the faint ripple of a telephone ring, unheard for three days, tinkled below. The children all sat up in bed to listen.

"What's that you say, Dave? MURDERED?" Everyone in the house was instantly awake. The girls leapt out of bed and lay down on the cold floor by the grate register to listen.

"The Chevalons? You don't say! I can't believe it!" Nobody breathed as they all strove to hear. "Haven't caught the murderer yet? EH?" A long silence followed. Then, "WIRE! Good heavens! AROUND THEIR NECKS? Mmmmm. Yes. We have two good guns and a twenty-two. Yeh. They are about four miles to the south. We're all right, Dave. And thanks for warning us." Sam hung up the phone. The girls above lay on the floor, paralyzed with cold and fright.

"For goodness sakes, Sam! You'll frighten the children —" Their voices lowered to hurried anxious whispers.

Teeth chattering, the children all crept back into bed to sleep fitfully all through the dreary night. At the slightest change in sound below one of them would leap up to peer out the window. Was that moving shadow across the snow the murderer skulking towards the house, or only a cloud passing over the moon? Were those sounds of splintering wood and dull thuds below someone breaking in to claim his next victim? Only a quick leap out of bed and a peek down the grate reassured them that it was their father replenishing the fire.

Morning came, clear and cold. With a jingle of sleighbells a floundering team dragging a cutter pulled into the yard. Mr. Wiley, the telephone repair man, tied his lathered horses to a post and stamped into the house to tell about the murder.

"Worst thing that ever happened in this district! A neighbour apparently discovered the tragedy when he was out looking for his cattle after the storm." Wiley took off his coat and accepted a steaming cup of tea.

"You girls get on in there and get the others ready for school. Late or not, Dad will take you today." Kate tried to keep the grisly details from the younger members of the family but it was a wasted effort. They all gathered around to listen,

"Windows were all smoked up. When he opened the door he saw where the bodies had fallen through a burnt hole in the floor and were piled up in a bloody heap in the cellar. Whoever did it certainly tried to burn the evidence. Oil on everything. Funny thing the fire went out, and in that wind, too. Blood all over the barn floor. Evidence shows they were killed there. At least Chevalon was. Must have dragged them by the neck to the house from the looks of things. It was over that girl. Body stripped of clothing, and covered with bruises. Chevalon must have fought to save them. The police came soon as they were called. Took some evidence away with them."

"The work of a brutish animal!" Kate wiped her eyes with the corner of her apron. "That poor, simple girl — her parents — as if life hadn't been hard enough already for them."

"Coroner was Dr. Hilts," continued Wiley. "He took young Dan McDougall and some others along to help; Archie Daley, Victor Mason and some of the LaMontagnes and neighbours. McDougall fainted the moment he laid eyes on the first corpse. Mason and some of the others hoisted the bodies out of the cellar and moved them outside with the wheelbarrow they had. Police are still working on the case and if it's anyone local it won't be long before they get him, I hope. Makes me nervous every time I have to leave the family." Wiley accepted another cup of tea.

"Funny thing though, Chevalon bought a lot of supplies from the store the other day but none of it can be found. Whoever did it got away with a lot of flour and stuff. Lowe says Chevalon spent over thirty-five dollars in the store. Biggest order he'd had for long enough so he remembers it well."

Edith thought of the sacks of flour and sugar the funny little man had tossed into his rig only a few short days ago, and shivered. Where were those sacks? Now she would never get to see the poor, simple girl who lived like a child and died like a woman.

The children climbed into the sleighbox and set off to school with dread for having to leave the safety of the house, even though their father was driving them. Recess and noons were filled with talk of the murders, for the next three days. Each evening Sam discussed the details with every neighbour over the phone. The nights were filled with nightmares, broken sleep and fear.

The third evening, as the children prepared for another sleepless night, the telephone rang shrilly.

"What?" Sam yelled. "What's that you say, Dave? They got him? They really did? Who? A long silence followed. Then, "VICTOR MASON? Archie told me they suspected him the other day. Mmmmm . . . . uhhuh . . . . Yes, we sure are! I'll just tell the wife."

Sam hung up the receiver and turned to Kate.

"It was that Mason fellow, all right. The French people in Dumas knew he was wanted in France for a murder there. The nerve of the brute. Mind you, he helped dig the bodies out of the cellar, too. Good for the police!"

Kate sat down and looked as if she were going to cry. She didn't even mention bedtime to the children.

"However did they find out it was he?"

"They found his mitts — ones Mrs. Chevalon had knitted for him, too — if that isn't irony. They were soaked with oil and covered with blood — his name was inside them — found them first thing but had to make sure. He forgot them, I guess, when he set fire to the place and then was afraid to go back for them. The police found all Chevalon's groceries at his place, too. Lowe identified them — remembered the boxes he put them in. Mason even stole one of those silver things they put on the top of coffins with the Chevalon name on it — he was using it for an ashtray. Glad it's over, I can tell you." Sam hugged his wife and smiled for the first time in days.

"Up to bed now, you kids. Get some sleep." He gave each an extra hug, and spent the next hour calling the neighbours and rehashing the affair.

The night winds around the house corners had different, friendly sounds. The children fell into bed and slept dreamlessly until morning.

CHAPTER 51

"Drat the luck! There's someone coming in the gate!" Kate put another plate on the table, and another kid shoved over on the bench. These complaints and the ensuing actions were a daily thing in her house. She rarely questioned the "rule of the plains" — "Feed all who come to your door at mealtime". This maxim demanded without saying it that a farm wife must give without question her fresh baked bread, her home-made pies and an open door hospitality, but required nothing from the recipient in return. There was no dearth of recipients at Kate's table.

The prairies were a man's world. Men poured across the land by the trainload; labourers, harvesters and homesteaders. They came from the "old country" and after a lifetime of seeding and reaping the prairie still spoke with the their Newcastle-on-Tyne drawl, their Yorkshire twang, or their monosyllabic middle European accent. Young marriageable teachers were always well-courted, and lost little time after arrival before they were wooed and carried off by the lucky men who won them. But many men were never quite up to grabbing the pretty prizes, or convincing them that a shack on an undeveloped prairie farm was heaven on earth.

To each farm family came these lonely ones who stopped as regularly as they had to pass by. The hunger in their eyes for human warmth was offset by the satisfied hunger after a meal at their neighbour's. Every Christmas, New Year's, Easter or Thanksgiving found one or more of these bachelors with their feet under the family table. They never left anything behind, like chocolates, perfume, or a gift for Kate; except if they stayed the night. Then often their unwelcome contributions had to be boiled and frozen out of the bedclothes, if Kate was lucky and caught them in time.

These men showed the lack of women in many ways. They lived by themselves, and their table manners showed it. Their cotton smocks were raggedy and out at the elbows. Holes in their grey woolen socks showed above their boot tops in summer, and their sheepskins always needed buttons in winter. Some never bathed, others never shaved and one (for whom Kate had to do the cooking for his threshers because he was too old) wrote mash notes to the older girls and proposed to a younger one when she was twelve.

Kate treated them all with the same cordiality she used with her best friends, and demanded the children do the same or she would "know the reason why!" when they were gone.

"I'm so tired of hired men, hired men, hired men! Just for once it would be nice to sit down to a meal with only the family around the table." Kate's wish was never to be granted. The hired man was as much a part of the prairie farm operation as the horses or machinery.

Some were adored, like Mike Hayes who helped keep the family alive during the bad flu epidemic, and who never forgot to bring back gum or presents when he went away. Others were tolerated, like the flies in the fall or the cold in the winter. Still others were, like the two Johns, to become part of the family, help educate and broaden the outlook of the children, and become neighbours and friends.

Stafiej, the Pole, spoke no English and couldn't read or write.

"He can pitch sheaves and handle horses. He's not stupid. I think I'll have him stay for winter chores," Sam confided to Kate. "I never need to yell at him."

The little man had a dignity which Sam respected, and over the years he became as much a part of the family as the dogs, cats, horses and kids.

"You get a scribbler now, and teach John how to read and write," instructed Kate. Edith was brand new at the skill herself, and avid to show off her learning to the little man who smiled at her efforts and said, "I ackey" to everything. The teacher's progress was miniscule until John Kirchenko, a Russian student from Manitoba University, arrived a short time later to help with the harvest.

"Mr. Kirchenko, would you like me to teach you to read?" Edith noted he had an accent, a sign he needed instruction.

He took the little girl gently on his knee, and she helped him sound out all the words in the primer. Her future career was determined the next day. Mr. Kirchenko picked up the newspaper, read the headlines, and discussed the news intelligently all evening with her father. Elated with success, she resolved to take education to the masses the moment she was grown.

John Kirchenko returned the next spring with his bride. They settled on the farm next door and for years shared all home celebrations with the Markses. They also shared their knowledge of the world and its culture with the family. The children learned of Man's inhumanity to Man through Mrs. Kirchenko's stories of Pogroms in the Jewish ghettoes in Russia. She often sang parts of operas, or gave descriptions of great stage artists to enliven many a cold winter's evening.

When John visited on winter evenings around the crackling heater, the children always got out of bed to listen at the grate. His deep, Russian-accented voice rolled throughout the household like martial music, sometimes unintelligible, but always exciting. He spoke often of his escape from Russia during the revolution, and of his experiences in the wilds of British Columbia.

"Vee come to a river and DEVIL but she was viillddd!" floated upstairs. The listeners, who had never known water any livelier than the spring thaw that burbled into the sloughs, or seen hills any higher than the top of the barn, could now hear winds that swept down from great mountain tops, and torrents carrying giant timbers crashing to the sea.

John Stafiej added another dimension to the children's lives. He brought home a litter of motherless wolves for pets. He helped semi-tame a nest full of crows. He showed the boys how to trap and prepare furs for market. All the children first learned how to drive a car in his Ford runabout, and he taught them how to play poker.

"You girls get the lamps cleaned!" It was an early winter night, and Kate was tired. "See you don't spill the oil, or it will taint the bread again. You'll need a good light for your homework."

After supper they gathered around the table. Kate sat by Sam near the gasoline lamp. She held a magazine close to her nose, putting on and taking off her glasses repeatedly as she struggled to read the printed pages. Sam read the newspapers.

"Listen to this, Missus." She looked up from her book, and he read of some world happening.

The rest of the family gathered at the other end of the table in a circle of yellow light shed from the coal oil lamps, school books open and pencils busy. Gradually homework came to a standstill, and they watched and listened as Buster and John talked quietly, dividing a box of wooden matches.

Kate peered down the table in the semi-gloom.

"What are you doing? Have you all got your homework done?"

"John's just teaching us how to play cards, is all."

"Mind it's only whist. Don't you gamble, now," and she turned her myopic eyes once more to her reading.

"We haven't any money to gamble with, Mama," and the matches passed quietly back and forth between them as they won or lost a bet.

John's aptitude at cards began to bother Kate. She heard rumours of gambling in a boxcar in town, and felt a moral obligation to warn him of its danger. This was a task only Kate would attempt, Since evidence of his success was indisputable.

One Sunday morning she said to Margaret, "Go up the back stairs and do the men's room before you get ready for Sunday School."

Margaret went upstairs and began to pick up discarded clothing. From John's suit, lying on the floor, cascaded a welter of dimes, nickels and quarters that jingled and rolled and glittered like all the jewels of Ophir, into every corner of the room.

"Mama! Mama! Come up here! Look at all this money!" Margaret shouted.

Kate resolved to do something about it when John came in from the chores.

"John," she began firmly, "gambling's a very bad thing." John bent over the tin wash basin.

"I ackey tink so, Missus." He sloshed more water on his face, and lots on the floor.

"It can get a hold on you." She scurried from stove to table. "It's as bad, if not worse, than liquor."

"I ackey tink you right, Missus. I ackey tink you right!" John agreed, lying happily and toweling most of the grime off onto the fresh roller towel.

He continued to look red-eyed and sleepless every Sunday morning, and the back stairs floor was usually spattered with change. Kate continued to tell him that gambling was a sin and would get a hold on him — until the depression struck in the early thirties. For some years money was so scarce around the home that John's gambling was never mentioned. His car was the only one with gas in the tank, and in it they went to dances, to get the mail, to school, and to church. During those hard times his gas tank, like the miraculous pitcher, was never empty.

"Go and get the horse and ride into town to get me some tea and things. Here's a list," said Kate. She gave Violet a handful of silver. "Just some money John gave me for doing his shirts," she mumbled.

"Oh, his gambling money . . .," Violet said innocently.

"Get on that horse, and hurry up about it." Kate cut off any other conversation, turned her back and began to thump and punch down a batch of bread.

On a warm Sunday morning years later, after the depression was over, John came in for breadfast, red-eyed and sleepy, with pockets that jingled.

"Gambling's a sin, John — worse than drinking liquor — you do well to think that over." Kate turned potatoes in the pan and broke in more eggs. "It can get a hold on you and ruin your life."

"I ackey tink so, Missus," agreed John, clearing his throat and making for the breakfast table. "I ackey tink you right."

The depression was over, and moral standards were once more back in perspective.

Sam didn't gamble, drink, or smoke, nor did he ever remark on those who did. His vocabulary contained such innocuous epithets as "beggar, ninny and jackass", which weren't considered profane. Occasionally he used the product of an unmarried union which, Kate explained, was Old English usage, not exactly swearing — but not to be used by any of the children.

At Christmas he cracked nuts for his offspring. He took them on his knee for their weekly nail-clipping, barbered their heads, and with his wife, tucked them in at night with a stealthy kiss on a sleeping cheek — all this after threshing and its anxieties were well past.

But he was a yeller. He yelled at men as well as at kids, and his voice got louder as threshing got closer, rippling across the ripe fields of grain like the commands of a nervy sergeant major.

He yelled loudest at Fritz, a courtly, middle-aged German, whose English was as sketchy as his knowledge of western farming. Fritz played the violin like Heifitz, listened intently to the high-schoolers discussing the League of Nations, and eagerly tried to improve his English, but he didn't know which end of a horse to hitch up, or how to put on a horse-collar.

Sam inherited Fritz from a neighbour who salted all sentences with cuss-words. Meals with Fritz were always interesting, as he tended to mix his learned profanity with his university learned English, often with startling results.

The girls and Kate enjoyed his music, but Sam was less than enthusiastic about the quality of service he gave.

A day or two after Fritz's arrival, the binder broke down. At noon Sam called Dan MacPherson, the implement dealer in town.

"Send the repairs home with the kids after school, will you, Dan?" he yelled. "I'll pull 'er apart and see what I can do until they get here!" He turned away from the phone.

"Fritz, you go and clean out the barn," and away off across the field Sam trotted to the broken down binder.

"Fritz," he hollered from across the pig pasture, "bring that crowbar, and come here and help."

All afternoon Sam's yelling continued. Fritz, puffing like a grounded flounder, trotted from barnyard to binder and back again, bringing what he thought Sam wanted, and taking it back again. The rest of the time he shoved and braced and lifted, and stood around while Sam disembowelled a total mystery of a machine.

Kate was busy in the kitchen baking bread when Fritz walked in, perspiring and confused.

"What is it, Fritz?" she asked, as she fished a pan of bread out of the oven.

"Madam," he said, in his most courtly manner, "are the bastards home from town yet?"

Kate's face turned red, then white. "What do you mean?" she asked, "what DO you mean?"

Just at that moment the democrat full of children drove into the yard.

"Aach! Yess! They are there!" Fritz bowed out, while Kate stood in shocked silence.

He rushed off, unloaded the pieces of iron from the rig, and walked back to the kitchen.

"Thank you, Madam, I haf now the bastards, which I am taking to your 'usbin'," he said, and shouldering the repairs, he made off across the field again.

CHAPTER 52

As the twenties ended, so too, did the family's happy way of life. By the thirties the winds of change had blown many good things away, along with the rich prairie soil that provided their living. The stock market crash of '29 didn't matter as much as the natural discomforts — the terrible winds, the dreadful lack of life-giving rain, and the hordes of grasshoppers which ate the grass to the roots and every other plant that dared to poke its head above the black prairie dirt.

The beginning of the thirties slowed everything to a near stop. Shadows that didn't go away with the next season began to gather over their heads. Although world wide prices for farm produce hit rock bottom, prairie farmers, the greatest gamblers in the world, sowed crops as usual. Shadows piled into sky-high clouds of dust, rolling across the flat land like giant vacuums. They sucked up every grain of soil from the pulverized fields and dropped them somewhere else when the wind died. The seed planted in one field blew across to another, grew for a time, and withered. Sloughs dried up, and only a trickle of water came from the wells.

Because of lack of feed and water, the herd of purebred Holstein cows, the horses and the pigs had to be slaughtered or sold for a pittance. Pieces of machinery, not yet paid for, lay scattered in all corners of the fields, and the silt sifted over and buried them. Balls of Russian thistle rolled and danced and tumbled over their dry graves. The land looked alien, and took on a different character.

Around the house the garden produced only those vegetables that were planted near the kitchen drain. The rest deteriorated under banks of swirling dust that sifted into the house, clouded the air everyone breathed, and gritted between their teeth from morning to night. The rows of trees Grandpa Marks had planted around the houseyard were almost bare of leaves, and their branches made mournful sounds as they rubbed against each other, creaking and swaying in the moaning winds.

Kate watched helplessly as Sam, like many other prairie farmers helpless against the overwhelming odds, became more and more depressed. He picked at the meagre fare at mealtimes and stared blankly, hour after hour, over his wasted fields.

One cold winter morning and hour or so before dawn, Kate stood dry-eyed and trembling by her son's bedside.

"Buster! Buster! Wake up," she cried. "Dad's disappeared. He's gone towards the tracks — no coat on. Hurry! The train's due in a little while!"

Kate, worn out by nights of disturbed sleep, watching over her husband, had fallen asleep herself and awakened to find him gone.

Buster grabbed his trousers, a sheepskin coat, and jammed bare feet into his overshoes.

"It's all right, Mama, he can't have gone far. I'll get him." He lit a lantern and put an arm around his mother's shaking shoulders.

"Don't worry, now. I'll reach him in time." Then Buster set off through the swirling drifts, following his father's tracks south. The black night closed in around. He floundered along, running as fast as he could and trying to judge the time left before the train was due. Several times he had to stop and go in a circle to pick up the trail of the half-obliterated footprints.

As the first pale light of morning began to melt the night shadows he could see the uncoated figure of his father, half-running, half-staggering towards the tracks.

"Thank God," he breathed. "Thank God," and he quickened his pace as the faint "whooo-ooo" of the approaching train sounded a few miles off.

Buster reached his father's side when he was only a few feet from the track, and put an arm gently around his shoulders as the roaring locomotive hurtled by. Sam, showing no surprise, turned large, sad eyes upon his son, then without a word turned and started homeward.

"You'll catch your death of cold, Dad. Here, you put on my coat." Buster helped his father into the sheepskin. "You sure move fast when you want to."

The faint flicker of a smile passed over Sam's face. "Glad you're here, son," he mumbled, and father and son, arm in arm, made their way slowly toward the house.

"We're going to sell out and move to B.C.," Kate announced to the family the following fall. Sam, after a period at the hospital, had recovered from a nervous breakdown. He had been packed off to the coast to spare him the agony of seeing all that was left of his life's work — a few wind-scarred buildings in the middle of drifts of dry black soil where nothing grew.

The boys had made another abortive attempt to produce something on the farm, but with no help from nature had given up the struggle.

One crisp October morning Kate, with eleven of her children, a Chesapeake dog and the family cat, packed up for the trip west. The children and the animals crawled into the back of the open truck. With Kate up front and Buster at the wheel they headed out for British Columbia. Tears filled the children's eyes and spilled over onto the robes and blankets piled around them in the back. They topped the first rise to the west and the house, yellow paint peeling, windows blank and lonely, dropped out of sight forever from their lives.

CHAPTER 53

Two weeks later signs along the highway indicated their new way of life was near at hand. The thing which changed their lives the most was water. Near the coast it began.

The children took off hats and turned their faces joyously to the weeping skies, letting the rain pour over their uncovered heads and down their necks, the first real heavy rainfall they had seen for several years.

Sam welcomed them to the place he had rented with an armful of late-blooming roses and an apple tree with apples still on it. The youngsters poured down out of the truck and scurried through the magic house, examining everything. It had hot and cold water taps throughout; a bathtub, toilet and bowl upstairs, and a sink for dishes with an extra washbowl in the kitchen.

No more struggle to carry water by the pailful up the hill, no fight to get men to bring up barrels for washing. Just a quick twist of the wrist and out flowed hot, hot water — soft water, no more thick grey scum on top. It was a miracle which they didn't begin to hate for several days.

A pretty little Jersey cow in her tiny stable at the back gave lots of rich milk.

"Come and see how I keep the milk cool," said Sam, and he led them down the cellar stairs. Lengths of tubing, remains of an old still someone had put in the auction barn where Sam spent every Saturday night, were rigged to a couple of tanks of water in which bottles of the Jersey's thick, cream-topped milk lay cooling. It took skill to make the bottles stay upright. The operation had to be scrapped for a whole milking if a bottle was dropped in carelessly, but it worked.

There were bowls of rich cream for porridge, and lots left over for churning, but the noise of the running water was clearly audible throughout the house, especially at night.

"Isn't rain-on-the-roof the loveliest sound in the whole world?" remarked Lorraine. The wide open upstairs had been curtained off into sleeping dormitories for the girls at one end, and the boys at the other. The first night, the rat-a-tat-tat on the shingles above was a reassuring sound to their prairie ears. They imagined thirsty plants soaking up moisture and promising abundant crops in the future.

"A field of prairie wheat would need water-wings to survive in this country," Margaret complained to Kate a week later.

On the prairies before the drought, clouds rolled up with the speed of a roaring freight train, blotting out the sun. Thunder rolled across the heavens, lightning ripped and the skies emptied welcome buckets of water on fields and gardens below. An hour later, the sun was back in business, shining and glittering and warm.

In the new home it didn't rain so much as it sogged — for weeks. Water oozed off the eaves, off the back porch and off the front verandah. It seeped through the shingles and dampened the beds upstairs. Every tap in the house plinked, plonked, dribbled and splashed. Lying sleepless at night and homesick for the prairies, the monotony of sounds had all the subtlety of the Chinese water torture. Everyone silently longed for a good, dry dirt-filled prairie wind.

Sam had adjusted to the climate, building up a trucking business, and he kept food on the table. He had no time for regrets. The rest of the family suffered in other ways.

In the classless society of the prairies, if you didn't have a job, but covered your poverty with courage and didn't steal, you were equal to anyone. The strong, wind-tanned personality of the plains was missing in the wet, vapid atmosphere of the coast. The family not only were poor, they felt poor. Besides that, Kate didn't like it.

"Drat this place," she snapped after a bad day, "We don't even know anyone here well enough to talk about them!"

There were painful things for the younger ones to learn, like the difference between prairie crocuses whose purple glory were free spring offering all the children made to their mother, and the sterile flowers cultivated in this coastal town.

Garnet bounded up the kitchen stairs from Sunday School on Mother's Day. He glowed with the uninhibited joy of childhood, and handed his mother a chubby fistful of yellow daffodils.

"Mmmmnnn, they're nice," Kate buried her face in the fragrant blooms.

"Where did you get them, Garnet?"

"Oh, I thust picked them up in front of the Empreth Hotel."

Kate explained his error to her son and apologized by telephone to the hotel manager for the loss of his border plants.

Time eased the aches. The rain let up a bit. Gardens awash with splashes of red tulips and golden daffodils made spirits soar. The gentle winds of spring brought the perfume of orchard blossomings, and they all got jobs.

CHAPTER 54

At the coast, with much of the family off their hands, Sam and Kate had time to do things they liked, which weren't always happily endorsed by their children.

Running a trucking business that specialized in hauling second-hand goods sold at auctions gave Sam lots of new ideas. Between his trucking trips he acquired a hobby. It was collecting things. His main surce of supply was Brown's Auction Barns on a Saturday night, where everything from freshening heifers to cracked gramophone records could be had for a price. And what prices! It was amazing what little outlay of money was required to fill the over-crowded house with wornout junk.

A stuffed mountain goat was the pride of his collection.

"It only cost one dollar and seventy-five cents," he would shout to groups of Sunday visitors gathered around the goat on the narrow front porch. It was always a shock to the girls of the family to note the delight on visitors' faces, but most of the people who came to call were misplaced stubble-jumpers too, who had never seen a mountain goat, much less owned one.

Some of the callers were not prairie bred. These were the eligible young men the unmarried daughters were hoping to attract.

Back home all young women had an equal chance. Unpainted houses, beat-up furniture or cracked crockery were no stumbling blocks to romance if you could smile, dance, or bake a good batch of bread. At the coast things were different. A sort of tacit social plateau prevailed which could never be attained by keeping a moth-eaten mountain goat on the front porch.

Sam seemed unaware of these values.

"Who moved the goat?" he yelled from the front yard.

"Just Saturday cleaning," Margaret lied from the back steps.

She and Hazel, caught in their weekly effort of trying to hide the thing, looked guilty.

"Not again," they muttered.

"Put it back now, mind, when you've done," Sam shouted cheerfully. "Back" meant shoving the monstrosity, the size of a small cow, onto the front porch in full view of every passing car, train or pedestrien. The stuffed brute, poised for jumping crags, stood like a king on a heavy iron pedestal.

It must be stuffed with iron, too," they grunted as they shoved and yanked and lifted to get it back where their father liked to show it off.

"He doesn't care if we all end up old maids," Hazel complained. They were expecting a young man who worked in a bank to come calling that evening.

"Every girl in town wants to go out with him," said Lorraine.

"No wonder. He's the only one that can take a girl out. None of the others have jobs. He's sure to like one of us — if it wasn't for this darn goat —"

"If we keep the front porch light off, and leave the one over the kitchen door on, perhaps he won't see it . . ." Violet looked hopeful.

There was little chance of success. They had tried on other occasions to get the goat moved to the back and to keep their father from missing it until after their dates were gone, but he seemed to unconsciously outwit them, making sure the callers saw his prize no matter where the girls put it. This always happened when the fellows were the slick-talking, smart-dressing, good dancers they favoured.

Alfred Watson, the bank clerk, was one of these. The girls had hardly manoeuvred him through the kitchen door before their father took over and manoeuvred him onto the dark front porch.

"Done any mountain climbing or big game hunting out here?" Sam boomed sociably, turning on the glaring front light.

Watson, who looked as if he'd never confronted anything gamier than the family cat, let out a startled bleat when he found himself nose-to-nose with the fierce-eyed goat.

"Look at that stance!" exclaimed Sam. "Look at that head! They can jump over chasms thousands of feet deep, and balance on rocks us humans couldn't get a toehold on!"

Watson must have suffered from acrophobia, for he turned pale, dashed for the safety of the front room, retrieved his hat, and with hasty excuses bolted out the back door.

"Funny fella," mused Sam, "doesn't seem to know much about his own province."

"You didn't give him much of a chance to say if he did," Kate remarked mildly. The girls, balked of their prey for a future date, slipped off to the bedroom to sulk.

The goat was the biggest cross to bear, but there were other painful slivers. Every Saturday night the back of the truck disgorged a welter of useless articles when the auction was over.

"We'll just stack these things here on the porch beside the goat until we find a proper place for them in the house," Sam would yell exultantly to whoever hadn't escaped to bed.

"Watch out for that Indian basket — the handles aren't so strong!" That was an understatement. Only one handle remained and it, like the basket, needed repair.

"Carry those dishes carefully now, so they won't get chipped." This, too, was an unnecessary caution. There wasn't a chipless one in the lot.

"Put this plant stand in the house and lean it against the wall until I find a leg for it." This gem, made of braided, twisted wickerwork, was impossible to clean and never did hold plants — who could find a braided, twisted wickerwork leg?

There were faded antimacassars, a rocking chair that humped as it rocked, a glass-doored gun cabinet with no glass, and beds . . .

Sam's bed period began insidiously. From the time of their arrival, he kept adding them until every room was jammed. As time passed, the girls began to show signs of revolt.

"Now don't you dare say a thing that will hurt Dad's feelings," Kate warned them all sharply, "he likes collecting things."

"But does he have to buy so darn many beds?" Kitt grumbled.

"Well," Kate said with resignation, for she was finding it difficult to move freely in the bed-packed rooms, "I guess it's because he always bought things in big quantities for so long, he can't get out of the habit. He doesn't realize there aren't so many of you home any more."

Sam bustled in from church. "Come and see what I bought last night, Vi." They went out through the back lean-to, a small shed built onto the kitchen to hold firewood. There was little room for wood in the tiny shed.

Every wall had bedsteads piled or leaning against it; small beds, large beds, wooden, steel, brass and curly iron ones — something to suit every size and shape of the human form.

Violet followed him down the path to the stable. It was full too, leaving very little room for hay or the pet Jersey cow. Against the outside wall leaned a real behemoth of a bedstead, nine feet tall, and almost as wide. It was solidly carved with curls and whorls and burls of wood. Naked noseless cherubs, wing-chipped, chased each other across the decorated panels. A wooden ribbon which strung them all together had great gaps missing, showing patches of unvarnished woodgrain beneath.

"Now what do you say to that?" he asked with pride.

Violet couldn't say they would never find a bed spring to fit it, or a place to put it, so she just looked and said nothing.

"It's a genuine antique — and it only cost me seventy-five cents!" He looked disappointed in her, and they went back to the house. Violet felt guilty all day, ashamed of letting him down.

The easy availability of books and magazines, and the great variety of fruit which flourished on the house lot, provided Kate with constant ammunition for her hobbies — reading and recipe experiments.

"I'm making Christmas wine," she explained one day when the kitchen was steamy with rich fruit odours, and glass jars, overflowing with juice, dripped over the table and onto the floor.

"

"The Liquor Board will be after you," Buster, just back from a logging camp, remarked slyly.

"This is NOT liquor. This is just Christmas wine!"

"If it gets alcohol in it, it's liquor."

"Nonsense! No such thing! There'll be no liquor in this house, now or ever! Just don't you forget that, young man. This is just Christmas wine — like preserved fruit, but only the juice!" And she went on happily throwing in this and that, boiling and bottling, and stashing it all in the cellar for the happy time ahead.

None of them ever got to drink it at Christmas. Long before the festive season arrived, the brew showed spirit of its own and blew the tops off all the bottles!

"Sure had a kick to that batch, Mama," teased Garnet. "Why don't you fix up a still — it would work better."

"Now, you can just stop that talk right now." Kate looked mad. "It's not liquor, and don't let me hear any of you say it is," she warned. "I just got the tops on a little crooked." She cleaned up the broken glass and opened all the windows to get rid of the smell of fermented mash.

Her experiments continued sporadically, but something always happened to the batches. That never deterred her. She liked trying, and she was content.

Time moved on. Some of the family married and made homes of their own, and Willie got religion.

"I'm going to see that every member of this family turns from evil ways!" he vowed to Kate one day.

"Nonsense!" she scoffed. "Get away with you. There's none of you smart enough to be evil!"

Near the end of Kate's life Hazel said to her, "Mama, weren't you surprised when we grew up with all our different personalities and pursuits?"

"Surprised?" Kate looked thoughtful. "I felt like a hen I had once, that hatched a clutch of duck eggs and then ran squawking her head off around the slough when they all began to swim!"

"Wasn't it awfully hard to leave Ontario and come out to the harsh climate and the primitive circumstances?" Robert asked.

Kate's blue, almost sightless, eyes had a soft, faraway look.

"It was glorious!" she sighed, "we were all young together."

Years later, Pat Marks, on a return visit to the old home, called on an aging neighbour.

"I thought your Dad was awful foolish in the twenties," he said, "He took you kids to all the picnics and dances and ball games round the country, and you all went flying by in your car to the lake. My family walked to town and did without. I saved, and promised myself I'd have money to give my kids a good education, so they enjoyed none of those things. When the depression hit, everyone went broke; your Dad, me, and all the rest. The thing is, you kids have all those happy memories."

Edith Hewson writes:

I have a special regard for all those friends and neighbours who were a part of the memories that helped create "Cornflakes" and shared the wonderful privilege of growing up on the prairies:

the Boyds, who watched us grow up;
the Wallaces, who put up with me so many nights;
the Davies, Lillian and her brothers who were teenagers with me;
the Kerchenkos, where I baby sat;
the Andrews, McClements, Pattersons, Toshs, Uptons, Wileys and Wests, who provided schoolmates;
and Mrs. Rev. Cairns, who taught us all to crochet.